Illustrations in Applied Network Theory

F. E. ROGERS, C. Eng., F.I.E.E., M.I.E.R.E.

Senior Lecturer, Department of Electrical and Electronic Engineering, The Polytechnic of Central London

LONDON
BUTTERWORTHS

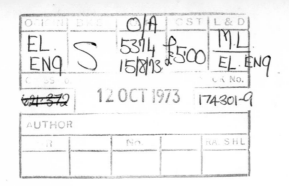
THE BUTTERWORTH GROUP

ENGLAND
Butterworth & Co (Publishers) Ltd
London: 88 Kingsway, WC2B 6AB

AUSTRALIA
Butterworths Pty Ltd
Sydney: 586 Pacific Highway, NSW 2067
Melbourne: 343 Little Collins Street, 3000
Brisbane: 240 Queen Street, 4000

CANADA
Butterworth & Co (Canada) Ltd
Toronto: 14 Curity Avenue, 374

NEW ZEALAND
Butterworths New Zealand Ltd
Wellington: 26-28 Waring Taylor Street, 1

SOUTH AFRICA
Butterworth & Co (South Africa) (Pty) Ltd
Durban: 152-154 Gale Street

First published in 1973

ISBN 0 408 70425 X Standard
     0 408 70426 8 Limp

Printed in Hungary

Illustrations in Applied
Network Theory

Preface

While this book is intended primarily for students in the intermediate years of a university or similar course, it includes many principles likely to be encountered in final year topics.

Books of worked examples have become popular, especially as guides to examination success. An examination, however, though useful as a target and difficult to replace as an impersonal but imperfect measure of attainment, is not the motive for education; and any book that facilitates examination success by imitation rather than by understanding contravenes not only the ideals of education but also the requirements of professional competence. On the other hand, understanding of principles can often be greatly enhanced by practical illustration. That is the motive for this book.

The division of teaching approaches into elementary and advanced is less dependent on stage in a given course than on type of course. In an academic course, the initial approaches should not be based on a provisional elementary outlook, but should embrace the mature premises on which later concepts are founded. In network and system theory, so-called elementary approaches have failed to impart an appreciation of duality and analogy, and have even resulted in the exclusion of nodal-voltage analysis on the pretext of difficulty. Yet its procedures are the duals of those for loop-current analysis, and are often easier to apply since branches linking nodes are generally discernible, whereas loops may often be obscure. Duality permeates this book, and nodal-voltage analysis, extended in a simple way to include the indefinite admittance matrix (not long ago thought too advanced a concept for early introduction), is used extensively.

While some examination questions have been included, most of the problems posed as illustrations have been designed specifically to highlight principles in an orderly way. Each one is given an interpretation that is intended to be more explicit than a mere solution, and is terminated with comment that often also collates the principles with other illustrations.

The author wishes to express his appreciation to his wife, for her encouragement and for typing the manuscript; to J. H. Gridley, Ph.D., B.Sc.(Eng), C.Eng., F.I.E.E., Head of the Department of Electrical and Electronic Engineering, The Polytechnic of Central London, for permission to use some questions from the C.N.A.A. Honours Degree course, and for helpful discussions on computer programming which, though not directly involved in this book, have influenced outlook; to Mr M. Martinho, B. Sc., for some manuscript reading; and especially to D. F. Neale, Ph.D., C.Eng., M.I.E.E., who has read much of the original manuscript and the proofs. Acknowledgement is also due to The Senate of The University of London for its kindness in permitting the inclusion of problems from its examination papers. The author alone is responsible for all solutions.

Hildenborough   F. E. ROGERS
Kent

Contents

CHAPTER 1

General principles for passive and active network analysis

INTRODUCTION

The purpose of this chapter is to illustrate the formulation and solution of network equations based directly on Kirchhoff's current and voltage laws, for a range of basic circuits of practical interest, especially in applied electronics. The principles illustrated are nevertheless common to all linear networks.

Competence in applying routine practical procedures for formulating network equations, such as nodal-voltage and loop-current analysis, requires a clear understanding of the evolution of such procedures from the initial premises of Kirchhoff's current and voltage laws. Accordingly, this introduction is devoted to an outline of this evolution.

In the illustrative network of Figure 1.1(a) the branches will be assumed, for simplicity, to comprise pure resistances or conductances, designated according to the branch numbering indicated. Kirchhoff's laws are concerned, not with the constituents of branches, but with the pattern of the network in the geometrical sense conveyed by Figure 1.1(b). Such a diagram, which defines the basic flow-paths and junction-points in a network, or its *topology*, is called a *graph*. In Figure 1.1(b) the graph is *oriented* with currents $j_1 \ldots j_6$, the directions of which are arbitrary; for directions are not only generally unpredictable, but are also of no consequence in establishing the equations: what is important is correlation of the *voltage falls* produced by them with the senses of action or *voltage rises* of sources in the network. It is advantageous to orientate branch voltages as falls in potential in the directions assumed for the currents, for then one oriented graph is common to both currents and voltages.

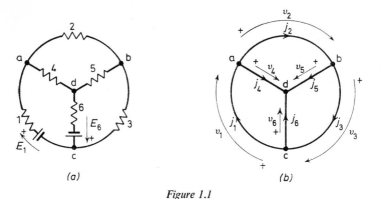

(a)   (b)

Figure 1.1

In this book an attempt is made to distinguish voltage rise from voltage fall by the attachment of a plus sign to the arrow-head when direction of rise is meant (as for E_1 and E_6 in Figure 1.1(a)); or to the arrow-tail, when direction of fall is meant (as in Figure 1.1(b)).

When there is an emf in series in a branch, its sense of action is the direction in which it would independently urge current, and this is regarded as its direction of voltage rise. For branch 1 in Figure 1.1(a), E_1 rises in the direction chosen for fall in Figure 1.1(b). Hence the net branch voltage is $v_1 = R_1 j_1 - E_1$. But for branch 6, the rise of E_6 is directed oppositely to j_6, and $v_6 = R_6 j_6 + E_6$. Dual (parallel) reasoning applies to the net current in the case of a current source acting in parallel with a branch. Both cases are generalised in Figure 1.2.

In terms of the net branch currents and voltages as implied in Figure 1.1(b), Kirchhoff's current and voltage laws (KCL, KVL) have the general forms

$$\sum_{k=1}^{n} j_k = 0 \tag{1}$$

and

$$\sum_{k=1}^{n} v_k = 0 \tag{2}$$

where n is the number of currents incident at a junction-point or *node* in equation (1), and the number of branch voltages in any closed circuit or *loop* in equation (2). Thus for Figure 1.1(b), taking currents arbitrarily as positive when directed towards a node,

$$j_1 - j_2 - j_4 = 0$$
$$j_2 - j_5 - j_3 = 0$$
$$j_3 - j_6 - j_1 = 0 \qquad (3)$$
$$j_4 + j_5 + j_6 = 0$$

while for the particular loops $(a\,d\,c\,a,\; a\,b\,d\,a,\; b\,c\,d\,b)$ that are the *meshes* (non-overlapping loops; as of a net; or 'windows', as suggested by E. A. Guillemin) of the network graph,

$$v_1 + v_4 - v_6 = 0$$
$$v_2 + v_5 - v_4 = 0 \qquad (4)$$
$$v_3 + v_6 - v_5 = 0$$

Meshes have been selected at this point rather than other possible loops because they facilitate the statement of loop-current as well as nodal voltage equations by inspection. It is important to realise, however, that while many practical networks may be *planar* and comprise distinct meshes as in Figure 1.1, many may have crossed branches, with a graph that is *non-planar* (formally, not mappable on a sphere), and meshes may not be discernible at all. In the general case, a formal topological approach is necessary for the systematic formulation of equations.

The removal of a certain minimum number of branches or *links* from a graph just opens all closed paths, leaving an open-branch structure called a *tree*. In Figure 1.1(b), removal of links 1, 2, 3 leaves the tree of branches 4, 5, 6. The restoration of one link at a time to this tree forms one *loop* or *tie-set* at a time, and each such loop is identifiable with one particular mesh of the complete graph. The tree of branches 4, 5, 6 is thus a special one out of many possible ones (16), in the sense that all its tie-sets are meshes. The removal of a different combination of these links, such as 2, 5, 6, also opens all closed paths; but the tree is a different one, and only that tie-set formed by restoring link 6 corresponds to a mesh.

In the formal topological approach to the general case, which is explained in some detail but in a simple way by the author in a companion, volume, 'Topology and Matrices in the Solution of Networks' (Butterworths 1965), tree-branch voltages are taken as the independent variables for the formulation of KCL equations, and link-currents for KVL equations. Thus, taking v_4, v_5 and v_6 as the independent variables for Figure 1.1(b), equations (4) gives

$$v_1 = v_6 - v_4, \quad v_2 = v_4 - v_5, \quad v_3 = v_5 - v_6 \qquad (5)$$

4

Then, defining the branches as conductances $G_1 \ldots G_6$ and allowing for the sources in Figure 1.1(a) in accordance with Figure 1.2(a),

$$j_1 = G_1(v_1+E_1) = G_1(v_6-v_4)+G_1E_1$$
$$j_2 = G_2v_2 = G_2(v_4-v_5), \quad j_3 = G_3v_3 = G_3(v_5-v_6) \tag{6}$$
$$j_4 = G_4v_4, \quad j_5 = G_5v_5, \quad j_6 = G_6v_6-G_6E_6$$

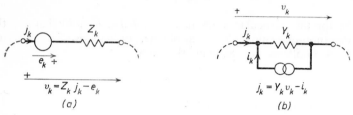

Figure 1.2. *The incorporation of sources into net branch voltage fall or net branch current*

Substituting these currents into the first three of equations (3) then gives

$$(G_1+G_2+G_4)v_4 \qquad -G_2\,v_5 \quad -G_1v_6 = G_1E_1$$
$$-G_2\,v_4+(G_2+G_3+G_5)v_5 \quad -G_3v_6 = 0 \tag{7}$$
$$-G_1\,v_4 \qquad -G_3\,v_5+(G_1+G_3+G_6)v_6 = G_6E_6-G_1E_1$$

In these equations the branch voltages v_4, v_5 and v_6 selected as independent variables are equivalently represented by the potentials of nodes a, b and c relative to the common node or *datum*, d; the emf E_1 acting in branch 1 of Figure 1.1(a) in the direction from node c to node a is replaced by an excitation current G_1E_1 flowing out of node c into node a; and similarly E_6 is replaced by a current G_6E_6 in the direction from node d to node c. These observations justify the equiv-

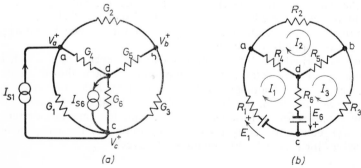

Figure 1.3

alent circuit representation of Figure 1.3(a), in which node d is taken as a datum to which assumed positive potentials for nodes a, b and c are referred, while E_1 and E_6, originally associated with series resistances, are replaced by current sources $I_{s1} = E_1/R_1 = G_1E_1$ and $I_{s6} = E_6/R_6 = G_6E_6$, acting in parallel with G_1 and G_6, respectively. The substitution of current sources in this way adapts the original circuit immediately to the dimensions of the current-law equations, and permits them to be written as nodal-voltage equations by inspection of the circuit.

In terms of Figure 1.3(a), in which the potential of the datum node d is taken as zero, the KCL equations may be written in full by inspection as

Node a: $\quad G_4V_a + G_1(V_a - V_c) + G_2(V_a - V_b) = I_{s1}$

Node b: $\quad G_5V_b + G_2(V_b - V_a) + G_3(V_b - V_c) = 0$ $\qquad$ (8)

Node c: $\quad G_6V_c + G_3(V_c - V_b) + G_1(V_c - V_a) = I_{s6} - I_{s1}$

On grouping terms these assume the same form as equations (7), or

$$(G_1+G_2+G_4)V_a \qquad -G_2 V_b \qquad -G_1 V_c = I_{s1}$$
$$-G_2 V_a + (G_2+G_3+G_5)V_b \qquad -G_3 V_c = 0 \qquad (9)$$
$$-G_1 V_a \qquad -G_3 V_b + (G_1+G_3+G_6)V_c = I_{s6} - I_{s1}$$

Equations (9) may be expressed more compactly in the ordered matrix style,

$$\begin{bmatrix} G_{11} & G_{12} & G_{13} \\ G_{21} & G_{22} & G_{23} \\ G_{31} & G_{32} & G_{33} \end{bmatrix} \cdot \begin{bmatrix} V_a \\ V_b \\ V_c \end{bmatrix} = \begin{bmatrix} I_{s1} \\ 0 \\ I_{s6} - I_{s1} \end{bmatrix} \qquad (10)$$

where $\quad G_{11} = G_1+G_2+G_4, \quad G_{22} = G_2+G_3+G_5, \quad G_{33} = G_1+G_3+G_6$

$\qquad G_{12} = G_{21} = -G_2, \quad G_{13} = G_{31} = -G_1, \quad G_{23} = G_{32} = -G_3$

and $\quad I_{s1} = G_1E_1, \qquad I_{s6} = G_6E_6$.

The alternative KVL set of equations is obtained by a parallel or dual procedure. The branches are defined as resistances; the independent variables are j_1, j_2, j_3, which are the currents in the links that form tie-sets or loops (in this case meshes) with the tree of branches 4, 5, 6; and the substitution is from equations (3) into (4), with $v_1 = R_1j_1 - E_1$ and $v_6 = R_6j_6 + E_6$. This procedure gives

$$(R_1+R_4+R_6)j_1 \qquad -R_4 j_2 \qquad -R_6 j_3 = E_1 + E_6$$
$$-R_4 j_1 + (R_2+R_4+R_5)j_2 \qquad -R_5 j_3 = 0 \qquad (11)$$
$$-R_6 j_1 \qquad -R_5 j_2 + (R_3+R_5+R_6)j_3 = -E_6$$

As in the KCL equations tree-branch voltages were identifiable with node to datum voltages, so in the KVL equations link currents are identifiable with loop (here mesh) currents as shown in Figure 1.3(b). Using the matrix notation, the KVL mesh equations can be written by inspection of Figure 1.3(b), in the form

$$\begin{bmatrix} R_{11} & R_{12} & R_{13} \\ R_{21} & R_{22} & R_{23} \\ R_{31} & R_{32} & R_{33} \end{bmatrix} \cdot \begin{bmatrix} I_1 \\ I_2 \\ I_3 \end{bmatrix} = \begin{bmatrix} E_1 + E_6 \\ 0 \\ -E_6 \end{bmatrix} \tag{12}$$

where $R_{11} = R_1 + R_4 + R_6$, $R_{22} = R_2 + R_4 + R_5$, $R_{33} = R_3 + R_5 + R_6$

$R_{12} = R_{21} = -R_4$, $R_{13} = R_{31} = -R_6$, $R_{23} = R_{32} = -R_5$

In this case no source transformations are involved as the sources (voltages) conform to the dimensions of the KVL equations.

Two relationships of parallel mathematical form, in which the one is changed into the other by a systematic interchange of coefficients and variables, are said to be *dual*. Consider, for example,

$$v = Ri, \quad v = L\frac{di}{dt}, \quad i = \frac{1}{C}\int i\,dt$$

$$i = Gv, \quad i = C\frac{dv}{dt}, \quad v = \frac{1}{L}\int v\,dt$$

These are dual relations, and establish duality between voltage and current, resistance and conductance (or more generally impedance and admittance), and inductance and capacitance. Duality is exploited in this book, and many practical illustrations will be found. The procedures of nodal voltage and mesh analysis outlined here are dual: comparing equations (9) with (11) shows that the sum of admittances meeting at a node has its dual in the sum of impedances forming the contour of a mesh, while an admittance linking two nodes has its dual in a mutual impedance common to two meshes. Topologically, a node may be shown to be the dual of a mesh, while the procedure of general loop-current analysis is precisely dual with the procedure of general node-pair voltage analysis.

The transformation from emf to current source that emerged in formulating the nodal voltage equations is general and reversible, provided the source is associated with a finite *immittance* (i.e., impedance or admittance). In equations (9) $I_{s1} = G_1 E_1$ is the short-circuit current of a generator whose emf is E_1 and series resistance is $R_1 = 1/G_1$. In the network this current acts in parallel with G_1; and as this arrangement must be equivalent to the original branch in which E_1 acted in series with R_1, it may be inferred in general terms that a generator of emf E and series impedance Z may be equivalently represented by a current $I = E/Z$ acting in parallel with $Y = 1/Z$; and conversely that a current I acting in parallel with Y may be equivalently represented by an emf $E = I/Y$ acting in series with an impedance $Z = 1/Y$ (note that this is an example of duality). In applying nodal-voltage and loop current analysis, these trivial transformations may be made at the start so that the sources may conform to the dimensions of the equations.

The nodal-voltage equations for Figure 1.1(a) may now be readily stated for any other node as datum. For example, if node b is the datum, V_b is set to zero in Figure 1.3(a) instead of V_d, while V_a, V_c and V_d are arbitrarly reckoned positive. Then, following the procedure exemplified by equations (9),

$$\begin{aligned}
(G_1+G_2+G_4)V_a \qquad &-G_1 V_c \qquad &-G_4 V_d = I_{s1} \\
-G_1 V_a + (G_1+G_3+G_6)V_c \qquad &-G_6 V_d = I_{s6} - I_{s1} \ (13) \\
-G_4 V_a \qquad &-G_6 V_c + (G_4+G_5+G_6)V_d = -I_{s6}
\end{aligned}$$

It is, however, valuable to refer the node potentials to an arbitrary point outside the network, for this generalises the equations so that the set for any particular node as datum is given by allowing this point to coincide with that node. The network under consideration has its own internal sources, and its nodes are, therefore, not linked to the external point through any closed-path. This raises a difficulty, for none of the node potentials has independent meaning in respect of an isolated point, although conjointly they must satisfy Kirchhoff's voltage law for all closed paths embraced by the nodes. This difficulty may be resolved through the contrivance of an emf between any one of the nodes and the external point, to support that node at any desired potential relative to the point: mathematically it is immaterial whether the potential of the node is attributed to this emf or to the network; but the substitution of the emf establishes a closed-path link between the network and the external datum point, and thus makes it a physically valid reference for the other node potentials. In applying the artifice to node b as indicated in Figure 1.4(a), the potentials of nodes a, c,

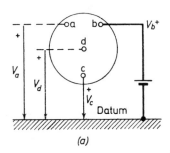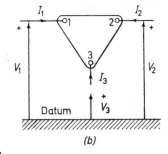

Figure 1.4

d relative to *b* are unchanged; but their falls relative to the external datum are all increased by the amount assumed for V_b: the nodes are 'floating' at a higher level than when *b* itself was taken as datum, as in equations(13).

For all nodes in Figure 1.3(a) assumed positive relative to an arbitrary external datum, the routine procedure gives

$$
\begin{aligned}
G_{11}V_a + G_{12}V_b + G_{13}V_c + G_{14}V_d &= I_{s1} \\
G_{21}V_a + G_{22}V_b + G_{23}V_c + G_{24}V_d &= 0 \\
G_{31}V_a + G_{32}V_b + G_{33}V_c + G_{34}V_d &= I_{s6} - I_{s1} \\
G_{41}V_a + G_{42}V_b + G_{43}V_c + G_{44}V_d &= -I_{s6}
\end{aligned}
\tag{14}
$$

or in matrix style,

$$
\begin{bmatrix}
G_{11} & G_{12} & G_{13} & G_{14} \\
G_{21} & G_{22} & G_{23} & G_{24} \\
G_{31} & G_{32} & G_{33} & G_{34} \\
G_{41} & G_{42} & G_{43} & G_{44}
\end{bmatrix}
\cdot
\begin{bmatrix}
V_a \\
V_b \\
V_c \\
V_d
\end{bmatrix}
=
\begin{bmatrix}
I_{s1} \\
0 \\
I_{s6} - I_{s1} \\
-I_{s6}
\end{bmatrix}
\tag{15}
$$

where $G_{11} = G_1 + G_2 + G_4$, $\quad G_{22} = G_2 + G_3 + G_5$, etc., and

$\qquad G_{12} = G_{21} = -G_2$, $\quad G_{23} = G_{32} = -G_3$, etc.

(a node admittance parameter, of general form Y_{kk} and on the leading diagonal of the matrix, is simply the sum of admittances meeting at node k; a transfer admittance parameter, of form Y_{ik}, is simply the negative of the admittance linking nodes i and k).

The conductance matrix in equation (15) is a form of *indefinite admittance matrix*, governing the behaviour of the network in the indefinite sense implied by an arbitrary datum, which precludes solution for absolute values of the voltages unless one at least is specified.

The set of equations for any particular node as datum is given by setting the potential of that node to zero. This is equivalent to allowing the arbitrary datum to coincide with that node. If, for example, node *b* is chosen as datum, putting $V_b = 0$ and removing the second equation from equations (14) gives, in matrix form,

$$
\begin{bmatrix}
G_{11} & G_{13} & G_{14} \\
G_{31} & G_{33} & G_{34} \\
G_{41} & G_{43} & G_{44}
\end{bmatrix}
\cdot
\begin{bmatrix}
V_a \\
V_c \\
V_d
\end{bmatrix}
=
\begin{bmatrix}
I_{s1} \\
I_{s6} - I_{s1} \\
-I_{s6}
\end{bmatrix}
\tag{16}
$$

On substituting for the parameters and expanding into normal equation form the result is the same as equations (13).

The conductance matrix in equations (16) is the *definite admittance matrix* for node *b* as datum, and is given directly by deleting from the indefinite matrix the row and column containing the node admittance parameter G_{22} identified with node b. Similarly, the definite matrix for node d as datum is given by deleting the row and column containing the element G_{44} which pertains to that node, giving

$$\begin{bmatrix} G_{11} & G_{12} & G_{13} \\ G_{21} & G_{22} & G_{23} \\ G_{31} & G_{32} & G_{33} \end{bmatrix} \cdot \begin{bmatrix} V_a \\ V_b \\ V_c \end{bmatrix} = \begin{bmatrix} I_{s1} \\ 0 \\ I_{s6} - I_{s1} \end{bmatrix} \qquad (17)$$

which may be compared with equation (10).

The indefinite admittance matrix exemplified in equation (15) has the very important property that the sum of its elements along any row or column is zero. For example, in the second column

$$G_{12} + G_{22} + G_{32} + G_{42} = -G_2 + G_2 + G_3 + G_5 - G_3 - G_5 = 0$$

This zero-sum property is the key to finding an indefinite matrix from a given definite one.

A general and more practical interpretation of the indefinite admittance matrix is given by Figure 1.4(b). The network, limited to three terminals or accessible nodes for simplicity only, is arbitrary. It is activated from external sources whose terminal voltages rise from a common datum, and the node currents are assumed to be in their positive senses. The assumption of external sources is not only practical, but also eliminates, through implied closed-paths, the difficulty attendant on an isolated datum. Note that external currents did not exist in the discussion of Figures 1.3 and 1.4(a) as the sources were internal and no closed paths existed to the external datum other than the fictitious one introduced to validate the concept.

The external behaviour of Figure 1.4(b) may be expressed in the general form

$$Y_{11}V_1 + Y_{12}V_2 + Y_{13}V_3 = I_1$$
$$Y_{21}V_1 + Y_{22}V_2 + Y_{23}V_3 = I_2 \qquad (18)$$
$$Y_{31}V_1 + Y_{32}V_2 + Y_{33}V_3 = I_3$$

Kirchhoff's current law must be obeyed for the network as a whole, or $I_1 + I_2 + I_3 = 0$. Adding equations (18) and imposing this condition gives

$$(Y_{11} + Y_{21} + Y_{31})V_1 + (Y_{12} + Y_{22} + Y_{32})V_2 + (Y_{13} + Y_{23} + Y_{33})V_3 = 0$$

But as the voltages are arbitrary, this can be true only if the coefficients, which are the sums of the elements in the columns of the indefinite admittance matrix, are zero. This confirms the zero-sum property more generally for columns, and also for rows in the case of a reciprocal network for which $Y_{ik} = Y_{ki}$. The property is not, however, restricted to such a network, and applies to a non-reciprocal device such as a transistor when operated under linear (small-signal) conditions (See Illustration 1.19). As a consequence of the zero-sum property, the determinant of the indefinite admittance matrix is zero. The matrix also has all its first cofactors equal, and is termed an *equicofactor matrix*.

By reference to equations (18), each element of the indefinite admittance matrix for a particular network may be found by imposing zero-voltage (short-circuit) constraints that exclude all elements other than the one of interest. For example,

$$Y_{11} = I_1/V_1 \Big|_{V_2 = 0, \ V_3 = 0}, \qquad Y_{32} = I_3/V_2 \Big|_{V_1 = 0, \ V_3 = 0}$$

(see Illustration 3.18).

For a network whose graph has B branches, N nodes and B_t branches to any tree formed by the removal of a minimum of L links,

$$L = B - N + 1, \quad B_t = B - L = N - 1$$

The minimum number of independent equations for solution of the network is L in the case of loop-current or mesh-current analysis, and B_t or $N-1$ in the case of nodal-voltage analysis. Thus one of these methods may be more economical than the other, according to the topology of the network. But this is not the only consideration. In a complicated network with crossed branches, meshes may not be discernible at all, and the formulation of KVL equations for other loops may require a formal topological approach for reliability. On the other hand, the nodes and their inter-linking branches are usually apparent even when the network is complicated and has crossed branches, so that nodal-voltage analysis is usually applicable in the simple routine way explained. When the network is likely to be used in alternative arrangements, as exemplified by a transistor in the common-base (CB), common-emitter (CE) or common-collector (CC) configurations, a knowledge of the indefinite admittance matrix is especially useful (see Illustration 1.19).

ILLUSTRATION 1.1

(1) Write by inspection the mesh equations for Figure 1.5(a) and the nodal-voltage equations for Figure 1.5(b). What inference can be drawn from the procedures?

(2) Write an alternative set of loop-current equations for Figure 1.5(a), choosing loops that are not all meshes; and an alternative set of nodal-voltage equations for Figure 1.5(b), choosing either node 2 or node 3 as datum.

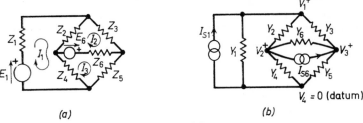

(a)   (b)

Figure 1.5

Interpretation

(1) The mesh equations for Figure 1.5(a) are

$$(Z_1+Z_2+Z_4)I_1 \qquad -Z_2 I_2 \qquad -Z_4 I_3 = E_1$$
$$-Z_2 I_1 + (Z_2+Z_3+Z_6)I_2 \qquad -Z_6 I_3 = -E_6 \qquad (1)$$
$$-Z_4 I_1 \qquad -Z_6 I_2 + (Z_4+Z_5+Z_6)I_3 = E_6$$

The assumption of identical (positive) signs for the node potentials in Figure 1.5(b) may be regarded as the dual of the assumption of identical (clockwise) directions of circulation of the mesh currents in Figure 1.5(a). In full, the KCL equations are

Node 1: $Y_1V_1+Y_2(V_1-V_2)+Y_3(V_1-V_3) = I_{s1}$

Node 2: $Y_2(V_2-V_1)+(Y_6(V_2-V_3)+Y_4V_2 = -I_{s6}$ $\qquad (2)$

Node 3: $Y_3(V_3-V_1)+Y_6(V_3-V_2)+Y_5V_3 = I_{s6}$

But these equations can be rearranged in the form

$$(Y_1+Y_2+Y_3)V_1 \qquad -Y_2 V_2 \qquad -Y_3 V_3 = I_{s1}$$
$$-Y_2 V_1 + (Y_2+Y_4+Y_6)V_2 \qquad -Y_6 V_3 = -I_{s6} \qquad (3)$$
$$-Y_3 V_1 \qquad -Y_6 V_2 + (Y_3+Y_5+Y_6)V_3 = I_{s6}$$

Comparing equations (3) with (1) shows that the sum of admittances at a node has its dual in the sum of impedances round the contour of a mesh, while an admittance linking two nodes is the dual of an impedance common to two meshes (a mutual impedance).

(2) Alternative equations may be formulated correctly for loops other than meshes by choosing loops that together include all branches of the network. This can be ensured by forming the loops from a tree of the network graph. Consider, for example, the tree of branches 1,6,5. The restoration, one at a time, of links 3,2,4 gives three loops which, when superimposed, restore the full structure of the graph.

Let currents I_1, I_2 and I_3 circulate in a clockwise direction (for convenience) in these loops. As the branch currents comprise the superimposed loop currents, it is necessary only to visualise the loops as superimposed in order to write the KVL equations

Loop 1,3,5: $(Z_1+Z_3+Z_5)I_1+(Z_1+Z_5)I_2+Z_5I_3 = E_1$

Loop 1,2,6,5: $(Z_1+Z_5)I_1+(Z_1+Z_2+Z_6+Z_5)I_2+(Z_5+Z_6)I_3 = E_1+E_6$

Loop 4,6,5: $Z_5I_1+(Z_6+Z_5)I_2+(Z_4+Z_6+Z_5)I_3 = E_6$

One of the loops only (4,6,5) is a mesh.

An alternative set of nodal-voltage equations for another datum can be written immediately by inspection. Let node 3 be chosen as datum so that $V_3 = 0$ while V_4 is now assumed finite and positive. Then,

$$
\begin{aligned}
(Y_1+Y_2+Y_3)V_1 && -Y_2V_2 && -Y_1V_4 &= I_{s1} \\
-Y_2V_1+(Y_2+Y_4+Y_6)V_2 && && -Y_4V_4 &= -I_{s6} \quad (5) \\
-Y_1V_1 && -Y_4V_2+(Y_1+Y_4+Y_5)V_4 &= -I_{s1}
\end{aligned}
$$

Comment

(1) When the network consists entirely of meshes there is generally no advantage in choosing alternative loops. Mesh equations can be written automatically but equations for alternative loops may not be obvious, as shown. Note the signs of E_6 in relation to the assumed directions of I_2 and I_3 in equations (1).

(2) The procedure of node-datum voltage analysis is dual with that of mesh-current analysis and is just as easy after the trivial transformation of emf sources with series impedances into current sources with shunt admittances has been made. Note that a current source appears positive when flowing into a node that has been assumed positive, as with I_{s1} in equations (3). Observe also that the nodal-voltage equations are equally easy to state for any datum node.

ILLUSTRATION 1.2

State Cramer's rule for the solution of an ordered set of equations, such as those pertaining to a linear network. Show how it yields compact expressions for the transfer and driving-point parameters of a network in terms of the determinant of the immittance-matrix and its cofactors.

For Figure 1.6 state compactly the transfer impedance $Z_{tr} = E_1/I_3$, the driving-point impedance $Z_{in} = E_1/I_1$ and the current ratio I_3/I_1.

Evaluate Z_{in} and I_3/I_1 at a scaled angular frequency $\omega = 2$ rad/s when the circuit elements have the scaled values $R = 1\,\Omega$, $C = 1$ F, $L = \frac{1}{2}$ H, $R_1 = \frac{1}{4}\,\Omega$ and $R_2 = 1\,\Omega$.

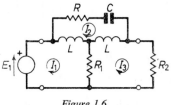

Figure 1.6

Interpretation

Consider a set of simultaneous equations in the matrix form

$$
\begin{bmatrix}
Z_{11} & Z_{12}\ldots Z_{1n} \\
Z_{21} & Z_{22}\ldots Z_{2n} \\
\ldots\ldots\ldots\ldots\ldots \\
Z_{n1} & Z_{n2}\ldots Z_{nn}
\end{bmatrix}
\cdot
\begin{bmatrix}
I_1 \\
I_2 \\
\vdots \\
I_n
\end{bmatrix}
=
\begin{bmatrix}
E_1 \\
E_2 \\
\vdots \\
E_n
\end{bmatrix}
\tag{1}
$$

Then Cramer's rule may be interpreted algebraically in the form

$$
I_k = (E_1\Delta_{1k} + E_2\Delta_{2k} + \ldots + E_i\Delta_{ik} + \ldots + E_n\Delta_{nk})/\Delta \tag{2}
$$

where Δ is the determinant of $[Z]$, or $\Delta = \det [Z]$, and Δ_{ik} is a typical cofactor of Δ, formed by deletion of the ith row and kth column from Δ. Its sign is positive when $i+k$ is even. Δ itself, of order n, may be resolved according to the Laplace development into n terms each of the form $Z_{ik}\Delta_{ik}$, where Δ_{ik} is a cofactor of order $n-1$.

For a network driven from a single emf such as E_1 in the first mesh,

$$
\frac{I_k}{I_m} = \frac{E_1\Delta_{1k}}{\Delta} \cdot \frac{\Delta}{E_1\Delta_{1m}} = \frac{\Delta_{1k}}{\Delta_{1m}} \tag{3}
$$

while a transfer impedance Z_{tr} relating E_1 to any current I_k has the generalised form

$$
Z_{tr} = E_1/I_k = \Delta/\Delta_{1k} \tag{4}
$$

The input or driving-point impedance is the special case for which $k = 1$ and

$$
Z_{in} = E_1/I_1 = \Delta/\Delta_{11} \tag{5}
$$

For Figure 1.6, the mesh equations are

$$\begin{bmatrix} Z_{11} & Z_{12} & Z_{13} \\ Z_{21} & Z_{22} & Z_{23} \\ Z_{31} & Z_{32} & Z_{33} \end{bmatrix} \cdot \begin{bmatrix} I_1 \\ I_2 \\ I_3 \end{bmatrix} = \begin{bmatrix} E_1 \\ 0 \\ 0 \end{bmatrix} \qquad (6)$$

where

$$Z_{11} = R_1 + j\omega L; \qquad Z_{22} = R + j(2\omega L - 1/\omega C); \; Z_{33} = R_1 + R_2 + j\omega L;$$
$$Z_{12} = Z_{21} = -j\omega L; \; Z_{13} = Z_{31} = -R_1; \qquad Z_{23} = Z_{32} = -j\omega L$$

Applying the Laplace development to $\Delta = \det[Z]$ along the first row, equations (4) and (5) give

$$Z_{tr} = E_1/I_3 = \Delta/\Delta_{13} = (Z_{11}\Delta_{11} + Z_{12}\Delta_{12} + Z_{13}\Delta_{13})/\Delta_{13} \qquad (7)$$

and similarly

$$Z_{in} = E_1/I_1 = \Delta/\Delta_{11} = (Z_{11}\Delta_{11} + Z_{12}\Delta_{12} + Z_{13}\Delta_{13})/\Delta_{11} \qquad (8)$$

where the easily evaluated second-order cofactors are

$$\Delta_{11} = Z_{22}Z_{33} - Z_{32}Z_{23}; \quad \Delta_{12} = -(Z_{21}Z_{33} - Z_{31}Z_{23});$$
$$\Delta_{13} = Z_{21}Z_{32} - Z_{31}Z_{22}$$

For the numerical values given,

$$\Delta_{11} = \tfrac{3}{4} + j\tfrac{23}{8}; \quad \Delta_{12} = -1 + j\tfrac{3}{2}; \quad \Delta_{13} = -\tfrac{3}{4} + j\tfrac{3}{8}$$

Then, substituting into equation (8) and simplifying,

$$Z_{in}(j2) = \frac{-8 + j19}{6 + j23} = 0 \cdot 69 + j0 \cdot 53 \; \Omega$$

The current ratio is simply the ratio of two cofactors. Thus, for $\omega = 2$,

$$\frac{I_3}{I_1} = \frac{\Delta_{13}}{\Delta_{11}} = \frac{-6 + j3}{6 + j23}$$

which may be simplified as desired.

Comment

Cramer's rule and the determinant of the immittance matrix are important especially for the systematization and generality they introduce into network relationships, as exemplified by equation (3) to (5).

Solution with the Laplace development is valuable, because it is automatic, for matrices not exceeding third order (3×3). For higher orders, however, its value is detracted from by the rate in growth of cofactors with order: the determinant of a 4×4 matrix, for example, expands into 12 second-order cofactors if it is not first simplified according to well-known properties of determinants. Often, however, Δ and its cofactors may have a purely symbolic role. An example is their use for establishing the concept of matrix inversion (see Illustration 1.20).

Note that in this illustration the input impedance could not be evaluated by compounding elements, unless the topology of the network is changed (see Illustration 3.18).

ILLUSTRATION 1.3

Use the determinant of the impedance matrix in equation (6), Illustration 1.2, to find a relationship between $Z_{11}\ldots Z_{33}$ satisfying the condition $I_3 = 0$ in Figure 1.6. Thence obtain expressions for R and C satisfying this condition at an angular frequency ω.

Interpretation

By Cramer's rule,

$$I_3 = E_1\Delta_{13}/\Delta = (Z_{21}Z_{32}-Z_{31}Z_{22})E_1/\Delta$$

where $\Delta = \det [Z]$ and Δ_{13} is the cofactor formed by deleting row 1 and column 3 from Δ. Thus, $I_3 = 0$ when

$$\Delta_{13} = Z_{21}Z_{32}-Z_{31}Z_{22} = 0 \tag{1}$$

where

$$Z_{21} = -(j\omega L), \quad Z_{32} = -(j\omega L),$$
$$Z_{31} = -R_1, \quad Z_{22} = R+j2\omega L-j/\omega C$$

Substituting in equation (1) then gives

$$-\omega^2L^2+R_1R-jR_1/\omega C+j2R_1\omega L = 0$$

whence, on putting the real and imaginary parts separately to zero,

$$R = \omega^2L^2/R_1 \quad \text{and} \quad C = 1/2\omega^2L$$

Comment

A null condition requires only that the cofactor should be zero, and evaluation of Δ is unnecessary. As any relevant cofactor for a three-mesh (or four-node) network is of second order only, its evaluation is simple.

ILLUSTRATION 1.4

Determine (1), the frequency at which the open-circuit voltage V_0 of the network in Figure 1.7 is in anti-phase with the input voltage V_i, and (2), the voltage ratio V_0/V_i at this frequency.

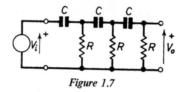

Figure 1.7

(L.U. Part 2, Electrical Theory and Measurements)

Interpretation

The problem may be approached in various ways. Of the fundamental methods, nodal analysis is less economical than mesh; but even this, if applied orthodoxly, could be tedious. However, the network is a uniform ladder structure, and this suggests an iterative approach.

The network is shown in general form in Figure 1.8, with V_0 reversed to represent the voltage fall in the sense of I_3.

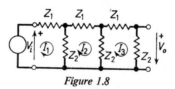

Figure 1.8

Let $Z_1/Z_2 = \lambda$, where λ is in general complex. Then, writing the KVL equations for each mesh in turn from right to left,

$$I_3(Z_1+2Z_2)-I_2Z_2 = 0$$

whence

$$I_2 = I_3(Z_1+2Z_2)/Z_2 = I_3(\lambda+2)$$

$$I_2(Z_1+2Z_2)-I_1Z_2-I_3Z_2 = 0$$

whence

$$I_1 = I_3(\lambda^2 + 4\lambda + 3)$$

$$I_1(Z_1 + Z_2) - I_2 Z_2 = V_i$$

or

$$I_3[(\lambda^2 + 4\lambda + 3)(Z_1 + Z_2) - (\lambda + 2)Z_2] = V_i$$

But $I_3 = V_0/Z_2$ and therefore

$$V_0[(\lambda^2 + 4\lambda + 3)(\lambda + 1) - (\lambda + 2)] = V_i$$

whence

$$\frac{V_0}{V_i} = \beta = \frac{1}{\lambda^3 + 5\lambda^2 + 6\lambda + 1}$$

In the case of Figure 1.7, $\lambda = 1/j\,\omega CR$
and

$$\frac{1}{\beta} = \frac{V_i}{V_0} = 1 - \frac{5}{(\omega CR)^2} + j\left[\frac{1}{(\omega CR)^3} - \frac{6}{\omega CR}\right]$$

When V_i and V_0 are in anti-phase, V_i/V_0 must be real and negative. The imaginary part is zero when

$$\frac{1}{(\omega CR)^3} - \frac{6}{\omega CR} = 0$$

whence

$$\omega = \frac{1}{\sqrt{6}\,C\,R}$$

Substituting this expression for ω into the real part then gives

$$\frac{1}{\beta} = \frac{V_i}{V_0} = 1 - 30 = -29$$

and

$$\left|\frac{V_0}{V_i}\right| = |\beta| = \frac{1}{29}$$

Comment

(1) Recognition of the iterative form of the network has permitted easy solution by progressive substitution: the formal solution of three simultaneous mesh equations has been circumvented.

(2) The circuit of Figure 1.7 is of practical importance as the basis of a common form of phase-shift oscillator. This is formed by returning the output of the ladder network to its input terminals through a

polarity-reversing amplifier, as indicated in Figure 1.9. When the gain of the amplifier is $A = -29$, the loop gain $A\beta$ at $\omega = 1/\sqrt{6}\,CR$ is just unity, and a steady oscillation at this frequency may just be sustained. Note that these conditions would be modified by loading effects if the amplifier input impedance could not be regarded as infinite, as in the case of a junction transistor.

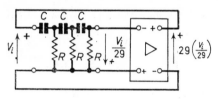

Figure 1.9

ILLUSTRATION 1.5

Figure 1.10 is the circuit of a common-emitter (CE) transistor amplifier operated under linear conditions for the amplification of small alternating voltages (of the order of a few mV). The circuit includes resistors (R_1, R_2, R_3) for biassing and stabilisation, and capacitors (C_1, C_2, C_3, C_4) for blocking dc and by-passing ac.

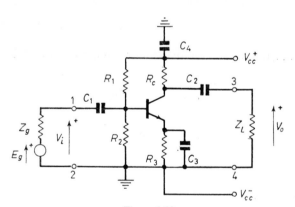

Figure 1.10

Draw a complete high-frequency equivalent circuit incorporating a hybrid-π circuit model for the transistor, and establish a set of nodal-voltage equations from which the gain V_0/V_i and other parameters might be found.

Interpretation

In respect of equivalent circuits, the classification of a frequency range as high or low is governed, broadly, by an assessment of the significance of series reactances and shunt susceptances. For Figure 1.10, a frequency is high when the intrinsic behaviour of the transistor is frequency-dependent; and this requires the transistor to be represented by an active circuit model such as the hybrid-π, in which the topological arrangement and values of capacitances and conductances in parallel give close simulation of the transistor's behaviour; from frequencies at which the capacitive susceptances are negligible, to frequencies, high in a relative sense, at which they are dominant. What is high, however, depends on the transistor: 50 MHz may be high for a radio-frequency transistor; but only 50 kHz may be high for an audio-frequency one. At frequencies high enough to justify consideration of the capacitors in the hybrid-π, the reactances of C_1 and C_2 would be negligible, and the susceptances of C_3 and C_4 would be so great that R_3 and the collector-emitter supply may both be regarded as short-circuited to ac. The only amplifier components for inclusion in the equivalent circuit are therefore R_1 and R_2, which form a parallel combination of conductance $G = (R_1+R_2)/R_1R_2$, and the collector load, of conductance $G_c = 1/R_c$.

Figure 1.11(a) is the complete equivalent circuit, with E_g, Z_g transformed into $I_g = E_g/Z_g = E_gY_g$ in parallel with Y_g, and Figure 1.11(b) is the circuit reduced to basic form in which $Y_1, \ldots, Y_5$ represent merged parallel branches (eg. $Y_5 = Y_L+G_c+g_{ce}+j\omega C_{ce}$) and the node

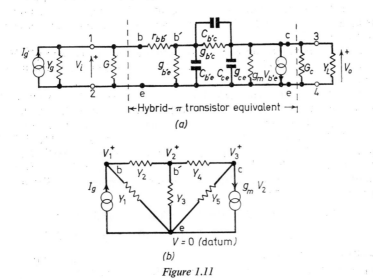

Figure 1.11

potentials, re-designated in numerical order for convenience and assumed positive relative to the datum, correspond respectively to V_i, $V_{b'e}$ and V_0.

By inspection of Figure 1.11(b),

$$
\begin{aligned}
(Y_1+Y_2)V_1 \quad\quad -Y_2\,V_2 \quad\quad\quad\quad &= I_g = E_g Y_g \\
-Y_2\,V_1+(Y_2+Y_3+Y_4)V_2 \quad -Y_4\,V_3 &= 0 \\
-Y_4\,V_2+(Y_4+Y_5)V_3 &= -g_m V_2
\end{aligned}
\tag{1}
$$

or in matrix style, combining g_m with Y_4 in the third equation,

$$
\begin{bmatrix}
Y_1+Y_2 & -Y_2 & 0 \\
-Y_2 & Y_2+Y_3+Y_4 & -Y_4 \\
0 & g_m-Y_4 & Y_4+Y_5
\end{bmatrix}
\cdot
\begin{bmatrix}
V_1 \\ V_2 \\ V_3
\end{bmatrix}
=
\begin{bmatrix}
I_g \\ 0 \\ 0
\end{bmatrix}
\tag{2}
$$

The gain V_0/V_i is simply the ratio of two cofactors of $\Delta = \det[Y]$; for $V_3 = I_g\Delta_{13}/\Delta$, $V_1 = I_g\Delta_{11}/\Delta$, and therefore

$$
\frac{V_0}{V_i} = \frac{V_3}{V_1} = \frac{\Delta_{13}}{\Delta_{11}} = \frac{-Y_2(g_m-Y_4)}{(Y_2+Y_3+Y_4)(Y_4+Y_5)+Y_4(g_m-Y_4)}
\tag{3}
$$

Comment

Note the ease with which the equations can be stated when the circuit is recast in basic form (topology). Similar procedure is applicable to a multistage amplifier with no increase in difficulty, and the resultant set of equations may be programmed in a high-level computer language such as Fortran IV (which is particularly suitable for complex quantities).

Other parameters are easily given from equations (1) and (2). For example, $Y_{in} = (I_g/V_1)-Y_g = (\Delta/\Delta_{11})-Y_g$ is the amplifier input admittance.

In practice, slide rule calculations may be simplified by making justifiable approximations. For example, a particular radio-frequency transistor, illustrative of the order of values in a hybrid-π equivalent-circuit, has: $g_{bb'} = 9.1\times10^{-3}$ S ($r_{bb'} = 110\ \Omega$); $g_{b'e} = 4\times10^{-4}$ S ($r_{b'e} = 2.5$ kΩ); $C_{b'e} = 400$ pF; $g_{b'c} = 5\times10^{-7}$ S ($r_{b'c} = 2$ MΩ); $C_{b'c} = 10.5$ pF; $g_m = 40$ mS; $g_{ce} = 4\times10^{-5}$ S ($r_{ce} = 25$ kΩ). C_{ce} is not quoted as it is small and inconsequential compared with the other parameters. At 5 MHz,

$$
\begin{aligned}
Y_4 &= g_{b'c}+j\omega C_{b'c} = (0.04+j3.3)\ 10^{-4} \simeq j3.3\times10^{-4} \\
Y_3 &= g_{b'e}+j\omega C_{b'e} = (0.04+j1.26)\ 10^{-2} \simeq j1.26\times10^{-2}
\end{aligned}
$$

while if Z_L is low (a few kΩ) and Y_L high, as in wide-band amplifiers, $Y_5 \simeq Y_L$ may also be justified; see also Illustration 3.16.

ILLUSTRATION 1.6

In Figure 1.12, $R_1 = R_2 = 1\ \Omega$, $L_1 = L_2 = 2$ H and $M = 1$ H. Calculate the input impedance at $\omega = 1$ rad/s:

(a), when L_1 and L_2 are orientated in accordance with the dot notation shown;

(b), when the winding direction of L_2 alone is reversed;

(c), when $R_1 = 0$ in both cases.

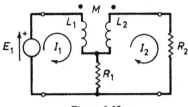

Figure 1.12

Interpretation

(a) For coil orientations implied by the dot notation shown, and for the mesh current directions assumed, I_1 enters L_1 by a dot but I_2 leaves L_2 by one. The self and mutual fluxes are therefore in opposition, and the mesh equations are

$$(R_1 + j\omega L_1)I_1 - j\omega M I_2 - R_1 I_2 = E_1$$
$$(R_1 + R_2 + j\omega L_2)I_2 - j\omega M I_1 - R_1 I_1 = 0$$

From these equations,

$$Z_{in} = \frac{E_1}{I_1} = R_1 + j\omega L_1 - \frac{(R_1 + j\omega M)^2}{R_1 + R_2 + j\omega L_2}$$

$$= 1 + j2 - \frac{(1 + j1)^2}{2(1 + j1)}$$

$$= \frac{1}{2} + j\frac{3}{2}\ \Omega$$

(b) When the winding direction of L_2 alone is reversed, the mutually induced voltages reverse to give additive self and mutual fluxes for the currents assumed in Figure 1.12. The dot notation, which indicates the ends of the coils at which assumed currents must both enter (or both leave) to produce additive fluxes, would thus demand transference of the dot on L_2 to the bottom of the coil. The mesh equations are changed in respect of the mutually induced voltages, which now become $+j\omega MI_2$ and $+j\omega MI_1$, and this leads to

$$Z_{in} = R_1 + j\omega L_1 - \frac{(R_1 - j\omega M)^2}{R_1 + R_2 + j\omega L_2}$$

$$= 1 + j2 + j(1 - j1)/2$$

$$= \frac{3}{2} + j\frac{5}{2}\ \Omega$$

(c) When $R_1 = 0$,

$$Z_{in} = j\omega L_1 - \frac{(\pm j\omega M)^2}{R_2 + j\omega L_2}$$

when $+j\omega M$ applies to case (a) and $-j\omega M$ to case (b). But

$$(\pm j\omega M)^2 = -\omega^2 M^2$$

so that for both cases,

$$Z_{in} = j\omega L_1 + \frac{\omega^2 M^2}{R_2 + j\omega L_2}$$

$$= j2 + (1 - j2)/5$$

$$= \frac{1}{5} + j\frac{8}{5}\ \Omega$$

Comment

It is important to note that, when the coupling between meshes of a network is mixed and includes mutual inductance, the driving point or input impedance depends on the sign of the mutual inductance. Only when the coupling is pure mutual inductance, as in case (c), is the input impedance independent of the winding senses of the coils. This, however, is not true of transfer parameters such as voltage or current ratios, or transfer impedance, since the polarity of the voltage induced in one coil by a current in the other depends on relative winding senses.

Whether the induced voltages aid or oppose the self-impedance voltages can be decided by examining the relative senses of the self and mutual magnetic fluxes, for given senses of current traversal. This is illustrated by reference to Figure 1.13 in which the interlinking or mutual fluxes ϕ_{21} and ϕ_{12} are proportions of ϕ_1 and ϕ_2, due to independent currents i_1 and i_2 respectively.

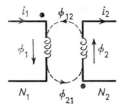

Figure 1.13

For the winding and current directions shown, both ϕ_1 and ϕ_2 are augmented by the mutual fluxes, so that the voltages across the coils of turns N_1 and N_2 become

$$v_1 = N_1 \frac{d}{dt}(\phi_1 + \phi_{12})$$

$$v_2 = N_2 \frac{d}{dt}(\phi_2 + \phi_{21})$$

(1)

or in terms of the self and mutual inductances,

$$v_1 = L_1 \frac{di_1}{dt} + M \frac{di_2}{dt}$$

$$v_2 = L_2 \frac{di_2}{dt} + M \frac{di_1}{dt}$$

(2)

For the sinusoidal steady-state, equations (2) become

$$V_1 = j\omega L_1 I_1 + j\omega M I_2$$
$$V_2 = j\omega L_2 I_1 + j\omega M I_2$$

(3)

Reversal of either winding or either current direction would reverse the mutual fluxes relative to the self-fluxes, thus reversing the signs of ϕ_{12} and ϕ_{21} in equations (1), and the sign of M in equations (2) and (3).

It is a convenient practice to mark with a dot, as indicated on Figure 1.13, those ends of the coil at which currents should enter or leave in order to produce self and mutual fluxes that are additive. It is useful to note that coil-ends marked with dots are of like polarity.

ILLUSTRATION 1.7

In Figure 1.14 the coils L_1 and L_2 coupled by mutual inductance M are marked with the dot notation, in accordance with the convention for coils coupled by mutual inductance.

(a) State the Kirchhoff voltage-law equations for both meshes, conforming to the emf and current orientations assumed in the diagram.

(b) Calculate the short-circuit transfer admittance I_2/E_1 at $\omega = 1$ rad/S when the network has scaled values $R_1 = R_2 = 1\,\Omega$, $L_1 = L_2 = 2$ H and $M = 1$ H.

(c) What admittance connected between terminals 1 and 3 would reduce the current traversing terminals 3 and 4 to zero? Synthesise a network to realise this admittance.

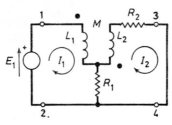

Figure 1.14

Interpretation

(a) The assumed mesh currents both enter the coils at the dotted ends. In accordance with the convention this implies that the self and mutual fluxes are additive, and the mesh equations are therefore

$$(R_1+j\omega L_1)I_1+j\omega MI_2-R_1I_1 = E_1$$
$$(R_1+R_2+j\omega L_2)I_2+j\omega MI_1-R_1I_1 = 0$$

(b) Substituting numerical values and manipulating gives

$$I_2 = \frac{E_1(1-j)}{-2+j8}$$

whence

$$Y_{21} = \frac{I_2}{E_1} = -\frac{5}{34}-j\frac{3}{34}$$

(c) An admittance Y connected across terminals 1–3 would produce an independent current E_1Y in the short-circuit, while that produced by the network of Figure 1.14 is E_1Y_{21}. The current in the short-circuit is therefore zero when $E_1Y+E_1Y_{21} = 0$, or when

$$Y = -Y_{21} = \frac{5}{34}+j\frac{3}{34}$$

For $\omega = 1$ rad/s, this admittance may be synthesised with a resistor $34/5\,\Omega$ in parallel with a capacitor of $3/34$ F.

Comment

(1) Note that the sign of Y_{21} depends on the directions assumed for the mesh currents: if I_2 were reversed, Y_{21} would reverse. While this is unimportant in the solution of an isolated problem, as in this illustration, it is important in the context of matrices; for representations in matrix form lose value if the forms are not universal and amenable to standard manipulations (see Illustrations 1.13, 1.14).

(2) The validity of summing short-circuit transfer admittances for a null condition can be seen in an alternative way illustrated by Figure 1.15. In this, Y_{21} is synthesised with an inductor of 34/3 H in parallel

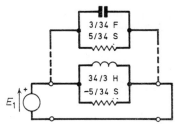

Figure 1.15. Equivalent representation for the transfer admittance of Figure 1.14.

with a negative conductance, $-5/34$ S. This admittance may be annulled at $\omega = 1$ rad/s by the parallel connection, as indicated, of a capacitor of 3/34 F and a physically realisable conductance of 5/34 S; for then

$$Y = \frac{5}{34} + j\frac{3\omega}{34} - \frac{5}{34} - j\frac{3}{34\omega}$$

$$= 0 \quad \text{when} \quad \omega = 1 \text{ rad/s}$$

(3) Note that the short-circuit transfer admittance alone gives only the short-circuit current of the network to which it relates, and is sufficient only for finding null conditions in the case of paralleled networks. It is insufficient to give correctly either the current in a finite load on an individual network, or on paralleled networks except for the special case of the null condition; see Illustrations 1.14, 1.15.

ILLUSTRATION 1.8

Explain the meaning of self inductance and mutual inductance and define them in terms of flux linkages and currents.

In the circuit shown in Figure 1.16, L_1 and L_2 are respectively the self inductances of the primary and secondary windings of an air-cored transformer, M being the mutual inductance between them. The winding sense from A to the point 0 is the same as that from B to the point 0.

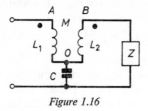

Figure 1.16

Find the nature and magnitude of the impedance Z that will ensure that points A and B will always be at the same potential when the excitation of the circuit is at an angular frequency $\omega = 10^5$ rad/s, and when $L_1 = 3{\cdot}2$ mH, $L_2 = 1$ mH, $M = 1{\cdot}2$ mH and $C = 0{\cdot}1$ μF. The windings have negligible resistance.

Find also the coupling coefficient between the coils.

(L.U. Part 2, Electrical Theory and Measurements)

Interpretation

(a) See Illustration 1.6, with particular reference to sign of M.

(b) The circuit is shown in a form suitable for analysis in Figure 1.17.

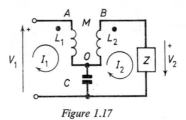

Figure 1.17

For the current directions assumed, I_1 enters L_1 by a dot but I_2 leaves L_2 by a dot, so that the mutually induced voltages oppose the self-induced voltages. In full, the KVL equations are therefore

$$\left(j\omega L_1 + \frac{1}{j\omega C}\right) I_1 - j\omega M I_2 - \frac{1}{j\omega C} I_2 = V_1$$

$$\left(Z + j\omega L_2 + \frac{1}{j\omega C}\right) I_2 - j\omega M I_1 - \frac{1}{j\omega C} I_1 = 0$$

and in compact matrix notation they have the form

$$\begin{bmatrix} Z_{11} & Z_{12} \\ Z_{21} & Z_{22} \end{bmatrix} \cdot \begin{bmatrix} I_1 \\ I_2 \end{bmatrix} = \begin{bmatrix} V_1 \\ 0 \end{bmatrix}$$

where

$$Z_{11} = j(\omega L_1 - 1/\omega C) = j2 \cdot 2 \times 10^2$$

$$Z_{22} = Z + j(\omega L_2 - 1/\omega C) = Z + j0 = Z$$

$$Z_{12} = Z_{21} = -j(\omega M - 1/\omega C) = -j20$$

The solution for I_2 is

$$I_2 = -\frac{V_1 Z_{21}}{Z_{11} Z_{22} - Z_{21} Z_{12}}$$

$$= \frac{j20 V_1}{j220 Z + 400}$$

The potential fall, $V_2 = I_2 Z$, in the sense of I_2, is required to equal V_1. Hence,

$$I_2 Z = V_1 = \frac{j20 V_1 Z}{400 + j220 Z}$$

or

$$4 + j2 \cdot 2Z = j0 \cdot 2Z$$

whence

$$Z = j2\Omega$$

In nature, Z is an inductor of value

$$L = \frac{Z}{j\omega} = \frac{2}{10^5} \, \text{H} = 20 \, \mu\text{H}$$

The coupling coefficient between the coils is simply

$$k = \frac{M}{\sqrt{L_1 L_2}} = 0 \cdot 67$$

Comment

This problem exemplifies the usefulness of the compact matrix notation. The manipulations might otherwise have been quite involved, for the simplicity of the result depends on the fact that L_2 is in resonance with C. This is revealed immediately Z_{22} is evaluated.

ILLUSTRATION 1.9

If the amplifier in Figure 1.18 has an input admittance Y in the absence of Y_f, derive an expression for the input admittance when Y_f is connected as shown, in terms of the ratio $A = V_2/V_1$. Adapt the result to the input admittance components of a triode valve amplifier giving a complex voltage amplification A, and having anode-grid and grid-cathode capacitances C_{ag} and C_{gk}, respectively.

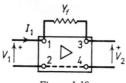

Figure 1.18

A parallel-tuned circuit, comprising an inductor of Q-factor 50 and inductance 500 μH in parallel with a capacitor that tunes it to resonance at $\omega = 10^6$ rad/s, is connected between the grid and cathode of a triode valve amplifier having an anode-grid capacitance of 3·5 pF and an effective voltage gain $A = 20\underline{/210°}$ at $\omega = 10^6$ rad/s. Calculate the effective Q-factor of the inductor at this angular frequency, and comment briefly on the effect of the valve on the tuning of the circuit.

(P.C.L., B Sc. (Hons.), C.N.A.A., Electronics, Year 2)

Interpretation

(a) For an intrinsic input admittance Y between terminals 1 and 2,

$$I_1 = YV_1 + Y_f(V_1 - V_2)$$
$$= YV_1 + Y_f(1 - A)V_1$$

The effective input admittance is therefore

$$Y_{in} = \frac{I_1}{V_1} = Y + Y_f(1 - A)$$

In the case of the triode valve, $Y = j\omega C_{gk}$, $Y_f = j\omega C_{ag}$ and

$$Y_{in} = j\omega C_{gk} + j\omega C_{ag}(1 - A)$$

Let the complex gain $A = |A|\underline{/\theta}$. Then

$$Y_{in} = j\omega C_{gk} + j\omega C_{ag}[1 - |A|(\cos\theta + j\sin\theta)]$$

From the real part,

$$G_{in} = \omega C_{ag} |A| \sin \theta$$

and from the imaginary part, jB,

$$C_{in} = jB/j\omega = C_{gk} + C_{ag}(1-|A|\cos \theta)$$

(b) Substituting the given values gives

$$G_{in} = -2 \cdot 5 \times 10^{-5} \text{ S}$$

while the shunt conductance of the coil is

$$G_L = 1/Q\omega L = 4 \times 10^{-5} \text{ S}$$

The net or effective conductance is therefore

$$G_{eff} = (4-2 \cdot 5) \ 10^{-5} = 1 \cdot 5 \times 10^{-5} \text{ S}$$

and the effective Q-factor of the coil is therefore

$$Q_{eff} = 1/\omega L G_{eff} = 133$$

The capacitance required to tune the coil to resonance at $\omega = 10^6$ rad/s is $C = 1/\omega^2 L = 2000$ pF. But the valve contributes

$$C_{in} = C_{gk} + 2 \cdot 5(1+20 \times 0 \cdot 866) = C_{gk} + 45 \cdot 8 \text{ pF}$$

While C_{gk} is not given, a reasonable value to assume would be about 4 pF, giving C_{in} as about 50 pF. The effective external tuning capacitance required would thus be about $2000 - 50 = 1950$ pF.

Comment

(1) The practical realisation of a negative conductance (or resistance) has been demonstrated. This is very important. As shown, the conductance of the coil has been substantially neutralised, and if the condition $G_L + G_{in} = 0$ was obtained, the tuned circuit would behave as if loss-free. A transient current (occasioned, for example, by switching on) could then establish a steady state of oscillation in the tuned circuit: this is the basis of the negative resistance (or, better, conductance) type of oscillator. The effect is detrimental in amplifiers, for it may cause instability to the point of oscillation, or at least impair the frequency response characteristics. A positive input conductance, while not contributing to instability, may damp the circuit connected to the amplifier input terminals and thus modify its behaviour. For a single-stage valve amplifier in the common-cathode circuit, or field

effect transistor in the common-source circuit, G_{in} may be negative when the anode (or drain) load is inductive. This places $\theta = \arg V_2/V_1$ in the third quadrant, for which $\sin \theta$ is negative.

(2) The effective input capacitance is much greater than the static partial capacitance C_{gk} (or C_{gs} in the case of an FET). This has the important implication that the effective capacitance shunting the amplifier source (which may be a pre-amplifier of high output impedance) is not only far greater than the apparant value C_{gk}, but tends to become proportional to A as A becomes great.

(3) The derivation of the expression $Y_{in} = Y + Y_f(1-A)$ is extremely simple when A is assumed (or measured in the presence of Y_f). Exact calculation without this assumption is more difficult. See Illustration 1.12.

ILLUSTRATION 1.10

The voltage gain of the amplifier bounded by terminals a, b and c, d in Figure 1.19 is A with Y_2 connected, while its input admittance at terminals a, b is zero when Y_2 is disconnected. Obtain an expression for V_2/V_1 and examine its tendencies for both positive and negative real values of A while Y_1 and Y_2 are treated algebraically as positive.

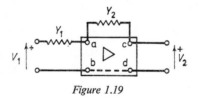

Figure 1.19

Let V_a denote the potential of node a, assumed positive relative to the common datum b...d. Then, for node a,

$$(Y_1+Y_2)V_a-Y_1V_1-Y_2V_2 = 0$$

Since $V_a = V_2/A$, this becomes

$$V_2[(Y_1+Y_2)/A-Y_2] = Y_1V_1$$

whence

$$\frac{V_2}{V_1} = \frac{AY_1}{Y_1+Y_2(1-A)}$$

Let ϕ denote a real, positive number. Then,
(1) When $A = -\phi$,

$$\frac{V_2}{V_1} = \frac{-\phi Y_1}{Y_1+Y_2(1+\phi)} = \frac{P}{Q}$$

P is always negative, while Q is always finite and positive, and

$$\frac{V_2}{V_1} \to -\frac{Y_1}{Y_2} \quad \text{as} \quad \phi \to \infty$$

(2) When $A = +\phi$,

$$\frac{V_2}{V_1} = \frac{\phi Y_1}{Y_1 + Y_2(1-\phi)} = \frac{P}{Q}$$

While P is always finite and positive, Q is finite and positive only for $\phi < 1 + (Y_1/Y_2)$, and is zero when $\phi = 1 + (Y_1/Y_2)$. As this condition is approached, $V_2/V_1 \to \infty$, and the system becomes unstable: at the threshold ($Q = 0$) a supposed excitation V_1 becomes self-sustaining, and the system may then act as a self-oscillator. The condition $\phi > 1 + (Y_1/Y_2)$, beyond the threshold of oscillation, cannot be treated as a linear inequality; for ϕ itself is then a function of oscillation amplitude, which it controls (with distortion) through the mechanism of amplifier non-linearity.

Comment

The circuit of Figure 1.19, used in the stable condition with A negative real and great in magnitude, is known as an *operational amplifier*: it is capable of performing integration and differentiation.

Let Y_1 be a conductance $1/R$ and Y_2 be the admittance of a capacitor C, and let the operations of differentiation and integration be denoted by D and 1/D, respectively. Then for a capacitor,

$$i(t) = C\frac{\mathrm{d}v(t)}{\mathrm{d}t} = C\mathrm{D}v(t)$$

so that the admittance Y_2 has the transient form

$$y_2(t) = i(t)/v(t) = C\mathrm{D}$$

Thus,

$$v_2(t) = -\frac{y_1(t)}{y_2(t)} \cdot v_1(t) = -\frac{1}{RC} \cdot \frac{v_1(t)}{\mathrm{D}} = -\frac{1}{RC}\int_0^t v_1(t)\,\mathrm{d}t$$

Conversely, when $y_1(t) = C\mathrm{D}$ and $y_2(t) = 1/R$,

$$v_2(t) = -RC\,\mathrm{D}v_1(t) = -RC\frac{\mathrm{d}}{\mathrm{d}t}v_1(t)$$

Alternatively, these properties are revealed instantly by Laplace transformation; for then,

$$V_2(s) = V_1(s) \cdot \frac{Y_1(s)}{Y_2(s)} = V_1(s) \cdot \frac{1}{sCR}$$

or

$$V_2(s) = V_1(s) \cdot sCR$$

where division by s and multiplication by s are recognised instantly as the transforms of integration and differentiation, respectively.

ILLUSTRATION 1.11

(a) Figure 1.20 is a circuit model, analogous to that for a triode electron tube, simulating closely a field-effect transistor (FET) when operated under substantially linear (small signal) conditions in the common-source circuit. Taking the source s as datum and assuming positive signs as indicated for both gate (g) and drain (d) potentials, write the nodal-voltage equations and thence deduce the admittance matrix for the FET.

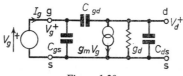

Figure 1.20

(b) An FET for which $g_m = 2 \cdot 0$ mS (mA/V), $r_d = 0 \cdot 5$ MΩ, $C_{gs} = C_{gd} = 5$ pF and C_{ds} is negligible, is used as a voltage amplifier in which an input voltage V_1 is applied between gate and source while the output voltage V_2 is developed across a drain-source load resistance of 10 kΩ. Find the voltage gain and phase change at $\omega = 2 \times 10^7$ rad/s.

Interpretation

(a) Let $j\omega = s$ and $1/r_d = g_d$. The nodal-voltage equations are then

$$s(C_{gs}+C_{gd})V_g \qquad\qquad -sC_{gd} V_d = I_g$$
$$-sC_{gd} V_g + [g_d + s(C_{ds}+C_{gd})]V_d = -g_m V_g$$

which may be rearranged in matrix form to display the admittance matrix for the FET in the common-source circuit as

$$\begin{bmatrix} Y_{11} & Y_{12} \\ Y_{21} & Y_{22} \end{bmatrix}_{cs} = \begin{bmatrix} s(C_{gs}+C_{gd}) & -sC_{gd} \\ g_m - sC_{gd} & g_d + s(C_{ds}+C_{gd}) \end{bmatrix}$$

(b) When a load Y_L is connected between drain and source, the matrix is modified by the addition of Y_L to Y_{22}. In the general matrix notation the nodal-voltage equations are then

$$Y_{11}V_1 + Y_{12}V_2 = I_1$$
$$Y_{21}V_1 + (Y_{22} + Y_L)V_2 = 0$$

and from the second,

$$\frac{V_2}{V_1} = -\frac{Y_{21}}{Y_{22} + Y_L} = -\frac{g_m - sC_{gd}}{Y_L + g_d + sC_{gd}}$$

$$= -\frac{(2 - j0\cdot1)10^{-3}}{(1\cdot02 + j1)10^{-4}}$$

$$= -14\underline{/-47\cdot3^\circ}$$

Comment

The matrix representing the transistor has been obtained simply by re-arranging the coefficients of the nodal-voltage equations in which positive signs have been assumed for the nodes in accordance with the practice already established. This assumption is, however, consistent with the convention for two-port analysis, and the matrix is therefore identical with that obtained through this approach in Illustration 1.16.

ILLUSTRATION 1.12

Calculate the input admittance components of the amplifier in Illustration 1.11; (a), from its Y-matrix; and (b), by using the known voltage gain in the expression $Y_{in} = Y + Y_f(1 - A)$ derived in Illustration 1.9.

Interpretation

(a) Let I_1 denote an external current source flowing from s to g and let V_1 and V_2 be the potentials of nodes g and d respectively, both assumed positive relative to s. Then, absorbing the load Y_L into Y_{22} as before,

$$Y_{11}V_1 \qquad + Y_{12}V_2 = I_1$$
$$Y_{21}V_1 + (Y_{22} + Y_L)V_2 = 0 \tag{1}$$

whence

$$Y_{in} = \frac{I_1}{V_1} = Y_{11} - \frac{Y_{12}Y_{21}}{Y_{22} + Y_L}$$

$$= s(C_{gs} + C_{gd}) + \frac{sC_{gd}(g_m - sC_{gd})}{g_d + sC_{gd} + Y_L} \tag{2}$$

Putting $s = j\omega = j2\times10^7$ rad/s and substituting for the FET parameters gives

$$Y_{in} = 1{\cdot}03 + j1{\cdot}15 \text{ mS}$$

The components are $G_{in} = 1{\cdot}03$ mS, and

$$C_{in} = j1{\cdot}15\times10^{-3}/j\omega = 57{\cdot}5 \text{ pF}$$

(b) Using the known gain of $-14/\overline{-47{\cdot}3^\circ}$,

$$\begin{aligned}
Y_{in} &= Y + Y_f(1 - A) \\
&= sC_{gs} + sC_{gd}(1 - A) \\
&= j10^{-4}[1 + (1 + 14/\overline{-47{\cdot}3^\circ})] \\
&= 1{\cdot}03 + j1{\cdot}15 \text{ mS}, \quad \overline{\text{as before.}}
\end{aligned}$$

Comment

The expression $Y_{in} = Y + Y_f(1 - A)$ is exact. From the second of equations (1), $A = V_2/V_1 = -Y_{21}/(Y_{22} + Y_L)$ so that equation (2) may be put in the form $Y_{in} = Y_{11} + AY_{12}$. But by the general 2-port principles developed in Illustration 1.13, $Y_{11} = Y + Y_f$ and $Y_{12} = -Y_f$. Therefore,

$$\begin{aligned}
Y_{in} = Y_{11} + AY_{12} &= Y + Y_f - AY_f \\
&= Y + Y_f(1 - A)
\end{aligned}$$

The simplicity of the expression makes it especially useful for the study of trends, such as dependence on A and arg A.

ILLUSTRATION 1.13

In Figure 1.21 (a) and (b) the currents I_1 and I_2 are directed as the assumed terminal voltages V_1 and V_2 woud independently urge them. Write the mesh equations for (a) and the nodal-voltage equations for (b) and restate them with ordered matrix coefficients. Then show that these coefficients may be defined equivalently by imposing open-circuit constraints on (a) and short-circuit constraints on (b).

Interpretation

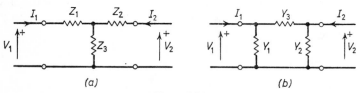

(a)   (b)

Figure 1.21

For (a), identifying mesh currents with I_1 and I_2,

$$\begin{aligned}
(Z_1 + Z_3)I_1 + Z_3 I_2 &= V_1 \\
(Z_2 + Z_3)I_2 + Z_3 I_1 &= V_2
\end{aligned}$$

or

$$Z_{11}I_1 + Z_{12}I_2 = V_1$$
$$Z_{21}I_1 + Z_{22}I_2 = V_2 \tag{1}$$

where

$$Z_{11} = Z_1 + Z_3, \quad Z_{22} = Z_2 + Z_3, \quad Z_{12} = Z_{21} = Z_3$$

For (b), identifying node potentials with V_1 and V_2,

$$(Y_1 + Y_3)V_1 - Y_3 V_2 = I_1$$
$$(Y_2 + Y_3)V_2 - Y_3 V_1 = I_2$$

or

$$Y_{11}V_1 + Y_{12}V_2 = I_1$$
$$Y_{21}V_1 + Y_{22}V_2 = I_2 \tag{2}$$

where

$$Y_{11} = Y_1 + Y_3, \quad Y_{22} = Y_2 + Y_3, \quad Y_{12} = Y_{21} = -Y_3$$

In equations (1), the constraint $I_2 = 0$ isolates Z_{11} and Z_{21}, while $I_1 = 0$ isolates Z_{22} and Z_{12}. The requirement that both voltages should exist while either of the currents is zero is satisfied physically by an open-circuit at either terminal pair; for then, though $I_2 = 0$, V_2 may exist as the open-circuit voltage due to V_1, I_1 alone; and similarly, when $I_1 = 0$, V_1 may exist as the open-circuit voltage due to V_2, I_2 alone. Thus,

$$Z_{11} = \frac{V_1}{I_1}\bigg|_{I_2=0} = \frac{I_1(Z_1 + Z_3)}{I_1} = Z_1 + Z_3$$

$$Z_{21} = \frac{V_2}{I_1}\bigg|_{I_2=0} = \frac{I_1 Z_3}{I_1} = Z_3$$

$$Z_{22} = \frac{V_2}{I_2}\bigg|_{I_1=0} = \frac{I_2(Z_2 + Z_3)}{I_2} = Z_2 + Z_3$$

$$Z_{12} = \frac{V_1}{I_2}\bigg|_{I_1=0} = \frac{I_2 Z_3}{I_2} = Z_3$$

Dual reasoning applies to equations (2) and Figure 1.21(b), for which short-circuits at either end satisfy the requirement that both I_1 and I_2 should exist while either V_1 or V_2 is zero. Thus,

$$Y_{11} = \frac{I_1}{V_1}\bigg|_{V_2=0} = \frac{V_1(Y_1 + Y_3)}{V_1} = Y_1 + Y_3$$

$$Y_{21} = \frac{I_2}{V_1}\bigg|_{V_2=0} = \frac{-V_1 Y_3}{V_1} = -Y_3$$

$$Y_{22} = \frac{I_2}{V_2}\bigg|_{V_1=0} = \frac{V_2(Y_2 + Y_3)}{V_2} = Y_2 + Y_3$$

$$Y_{12} = \frac{I_1}{V_2}\bigg|_{V_2=0} = \frac{-V_2 Y_3}{V_2} = -Y_3$$

Comment

(a) Note that Y_{12} and Y_{21} are negative for the polarity convention used, since I_2 and I_1 are opposite to the senses of action of V_1 and V_2 respectively.

(b) The matrix for a network may be regarded as a substitute for its circuit, so that the external behaviour can be fully determined without reference to the internal circuitry. In the case of the T network, for example,

$$[Z] = \begin{bmatrix} Z_{11} & Z_{12} \\ Z_{21} & Z_{22} \end{bmatrix} = \begin{bmatrix} Z_1 + Z_3 & Z_3 \\ Z_3 & Z_2 + Z_3 \end{bmatrix}$$

and in matrix style equations (1) are written as

$$\begin{bmatrix} Z_{11} & Z_{12} \\ Z_{21} & Z_{22} \end{bmatrix} \cdot \begin{bmatrix} I_1 \\ I_2 \end{bmatrix} = \begin{bmatrix} V_1 \\ V_2 \end{bmatrix}$$

where V_1 and V_2 are governed by the form of the external circuitry.

(c) While the Z and Y matrices for simple passive, reciprocal ($Z_{12} = Z_{21}, Y_{12} = Y_{21}$) structures may often be obvious from observation of the impedances or admittances under the requisite open or short-circuit conditions, the fundamental procedure of imposing constraints on voltages and/or currents is essential in the case of non-reciprocal devices such as transistors.

(d) The current and voltage orientations shown in Figure 1.21 are customary for the specification of two-port networks by matrices, in order to ensure standardisation of matrix forms and transformations, which is essential for their ordered application to groups of networks. Transformations from one form to another vary in signs according to the convention used; and two-port matrix representations lose value if the forms are not universal and amenable to standard manipulations. This does not mean, however, that the analysis of a two-port circuit in isolation, in which coefficients are organised in matrix form for convenience, must be bound to this convention (see, for example, Illustrations 1.2, 1.8).

ILLUSTRATION 1.14

The network N is defined equivalently by

$$[Z] = \begin{bmatrix} Z_{11} & Z_{12} \\ Z_{21} & Z_{22} \end{bmatrix} \quad \text{and} \quad [Y] = \begin{bmatrix} Y_{11} & Y_{12} \\ Y_{21} & Y_{22} \end{bmatrix}$$

(1) Write the equations for the whole circuit in terms of each matrix.
(2) Find expressions for I_2/I_1 and V_2/V_1 by selecting the optimum equations.

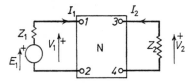

Figure 1.22

(3) Calculate I_2/I_1, V_2/V_1, V_2/E_1 and the input impedance Z_{in} or admittance Y_{in} at terminals 1, 2 for a purely resistive network and external circuit, when $Z_1 = 5\ \Omega$, $Z_2 = 10\ \Omega$, $Z_{11} = 10\ \Omega$, $Z_{12} = Z_{21} = 5\ \Omega$, $Z_{22} = 15\ \Omega$ and $Y_{11} = 0{\cdot}12$ S, $Y_{12} = Y_{21} = -0{\cdot}04$ S, $Y_{22} = 0{\cdot}08$ S.

Interpretation

(1) The Z-matrix equation is formally

$$\begin{bmatrix} Z_{11} & Z_{12} \\ Z_{21} & Z_{22} \end{bmatrix} \cdot \begin{bmatrix} I_1 \\ I_2 \end{bmatrix} = \begin{bmatrix} V_1 \\ V_2 \end{bmatrix} = \begin{bmatrix} E_1 - Z_1 I_1 \\ -Z_2 I_2 \end{bmatrix} \tag{1}$$

and the corresponding mesh equations are

$$Z_{11}I_1 + Z_{12}I_2 = V_1 = E_1 - Z_1 I_1 \tag{2a}$$

$$Z_{21}I_1 + Z_{22}I_2 = V_2 = -Z_2 I_2 \tag{2b}$$

As $[Y]$ conforms to current sources and nodal voltage equations, it is expedient to transform the emf generator E_1, Z_1 into an equivalent current generator comprising a current source $(E_1 Y_1)$ in parallel with Y_1, where $Y_1 = 1/Z_1$, and to replace Z_2 by $Y_2 = 1/Z_2$. The nodal voltage equations are then

$$Y_{11}V_1 + Y_{12}V_2 = I_1 = (E_1 Y_1) - Y_1 V_1 \tag{3a}$$

$$Y_{21}V_1 + Y_{22}V_2 = I_2 = -Y_2 V_2 \tag{3b}$$

(2) The current ratio is given directly from equation (2b) and the voltage ratio from equation (3b). From (2b),

$$Z_{21}I_1 + (Z_{22} + Z_2)I_2 = 0$$

whence

$$I_2/I_1 = -Z_{21}/(Z_{22} + Z_2) \tag{4}$$

Similarly, from (3b),

$$V_2/V_1 = -Y_{21}/(Y_{22} + Y_2) \tag{5}$$

(3) Substituting numerical values in equation (4) and (5) gives

$$I_2/I_1 = -1/5 \text{ and } V_2/V_1 = 2/9$$

Either equations (2) or (3) will yield V_2/E_1 and $Z_{in} = V_1/I_1$ or $Y_{in} = I_1/V_1$, but equation (3a) yields both easily. Substituting $V_1 = 4 \cdot 5 V_2$,

$$V_2/E_1 = Y_1/[4 \cdot 5(Y_{11}+Y_1)+Y_{12}] = 1/7$$

and substituting $V_2 = V_1/4 \cdot 5$ gives

$$Y_{in} = I_1/V_1 = Y_{11}+Y_{12}/4 \cdot 5 = 0 \cdot 111 \text{ S}$$

Comment

(a) The right-hand sides of equations (1), (2) and (3) are the adaptions of the general two-port equations to the particular external circuitry. Observe that $I_1 = (E_1 Y_1) - Y_1 V_1$ is the dual of $V_1 = E_1 - Z_1 I_1$, and that as V_2 is a voltage fall due to I_2 traversing Z_2, $V_2 = -I_2 Z_2$ or $I_2 = -Y_2 V_2$. These substitutions are automatic for a terminated network.

(b) While equations (2) and (3) are equally flexible, the directness with which the derived expressions, equations (4) and (5), express respectively the current and voltage ratios is important. The negative current ratio $(-1/5)$ means that the true direction of I_2 is opposite to that assumed in Figure 1.22. This is consistent with the positive voltage ratio, however, since V_2 is actually a voltage fall from terminal 3 to terminal 4.

(c) Equation (5) arises often in amplifier calculations, for it expresses the voltage gain simply and is adaptable to parallel feedback arrangements (see Illustrations 1.11, 1.16).

ILLUSTRATION 1.15

Two two-port networks, each defined by a Y-matrix, are connected in parallel. Deduce the conditions under which the Y-matrix for the composite arrangement is the sum of the Y-matrices for its parts.

Interpretation

Let the independent sets of equations corresponding to the individual Y-matrices be

$$Y'_{11}V'_1+Y'_{12}V'_2 = I'_1$$
$$Y'_{21}V'_1+Y'_{22}V'_2 = I_2$$

and

$$Y''_{11}V''_1 + Y''_{12}V''_2 = I''_1$$
$$Y''_{21}V''_1 + Y''_{22}V''_2 = I''_2$$

when the networks are connected in parallel so that $V'_1 = V''_1 = V_1$ and $V'_2 = V''_2 = V_2$, the composite behaviour is expressed by the addition of corresponding equations, provided that the currents in the one set remain independent of those in the other. Under this condition,

$$(Y'_{11}+Y''_{11})V_1 + (Y'_{12}+Y''_{12})V_2 = I'_1+I''_1 = I_1$$
$$(Y'_{21}+Y''_{21})V_1 + (Y'_{22}+Y''_{22})V_2 = I'_2+I''_2 = I_2$$

or

$$\begin{bmatrix} Y'_{11}+Y''_{11} & Y'_{12}+Y''_{12} \\ Y'_{21}+Y''_{21} & Y'_{22}+Y''_{22} \end{bmatrix} \cdot \begin{bmatrix} V_1 \\ V_2 \end{bmatrix} = \begin{bmatrix} I_1 \\ I_2 \end{bmatrix}$$

and

$$[Y]_{\text{system}} = [Y]' + [Y]''$$

When the networks are paralleled, the currents I'_1, I''_1 and I'_2, I''_2 can be independent of each other only if the volt-ampere relations within each network are not disturbed from the proportions set by the intrinsic Y-parameters. These are independently determined by imposing short-circuit conditions consistent with the constraints $V'_1 = 0$, $V'_2 = 0$ and $V''_1 = 0$, $V''_2 = 0$.

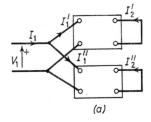 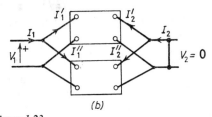

(a)   (b)

Figure 1.23

In Figure 1.23(a) the two networks are shown supplied from a common voltage source V_1 while the output terminals are separately short-circuited. The conditions in Figure 1.23(a) under which Y'_{11}, Y'_{21} and Y''_{11}, Y''_{21} are defined independently must be preserved when a common short-circuit is applied to the paralled networks as in Figure 1.23(b). Thus, for example, it is required that

$$Y_{21} = \frac{I_2}{V_1} = \frac{I'_2}{V_1} + \frac{I''_2}{V_1} = Y'_{21} + Y''_{21}$$

But this can be true only if there is no potential difference between the short-circuits, so that if they are linked by a conductor no current would flow in this link to disturb I_2' and I_2'': only then can the common short-circuit in Figure 1.23(b) carry a current I_2 that is the undisturbed superposition of I_2' upon I_2''.

Comment

(a) The validity test must be applied to both sides of the networks. However, Y-matrix addition is generally valid for networks of the unbalanced-to-ground type, which are really three-terminal networks having one terminal duplicated to provide distinct input and output terminal-pairs. Examples are the T-networks in in Figure 1.24(a), and it is obvious that the test $V = 0$ is satisfied by them. T-networks are of particular interest for paralleling.

(b) For networks connected in series so that I_1 and I_2 are to be common to them, the conditions are the duals of those for paralleled networks and

$$[Z]_{\text{system}} = [Z]' + [Z]''$$

provided the criterion indicated in Figure 1.24(b) is satisfied for both sides.

(c) In the hybrid equations

$$h_{11}I_1 + h_{12}V_2 = V_1$$
$$h_{21}I_1 + h_{22}V_2 = I_2$$

h_{11} is the input impedance under a short-circuit condition at the output port ($V_2 = 0$), while h_{22} is the output admittance under an open-circuit condition at the input port ($I_1 = 0$). This suggests the applicability of h-matrices to networks series-connected at input ports and parallel connected at output ports. Provided the open-circuit criterion for series connection is satisfied at the input ports when the paralleled output ports are energised by V_2, and the short-circuit criterion is satisfied at the output ports when the series connected input ports are energised by I_1,

$$[h]_{\text{system}} = [h]' + [h]''$$

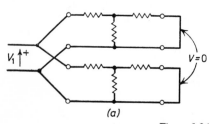
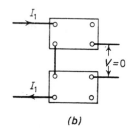

(a)   (b)

Figure 1.24

(d) It is often necessary to incorporate isolating transformers to validate the network interconnections. See Illustration 6.5.

ILLUSTRATION 1.16

A field-effect transistor in the common source circuit has a mutual conductance g_m, a differential drain resistance r_d, and inter-electrode capacitances C_{gs} (gate-source), C_{gd} (gate-drain) and C_{ds} (drain-source). Assuming linear operating conditions, show that the admittance-matrix is

$$[Y] = \begin{bmatrix} s(C_{gs}+C_{gd}) & -sC_{gd} \\ g_m-sC_{gd} & g_d+s(C_{ds}+C_{gd}) \end{bmatrix}$$

A field-effect transistor for which $g_m = 2{\cdot}0$ mA/V, $r_d = 0{\cdot}5$ MΩ, $C_{gd} = 5{\cdot}0$ pF, and C_{ds} is negligible, is used as a voltage amplifier with a drain-source load resistance of 10 kΩ. Calculate the voltage gain and phase-change at an angular frequency $\omega = 2 \times 10^7$ rad/s, with reference to an input voltage applied between gate and source.

(P.C.L., BSc. Hons. (C.N.A.A.), Electronics, Year 2)

Interpretation

(a) The transistor equivalent circuit is shown in Figure 1.25, with external voltages and currents oriented in accordance with the convention for 2-port network analysis and r_d replaced by $g_d = 1/r_d$.

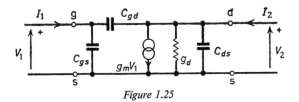

Figure 1.25

The admittance matrix may be found by imposing the constraints $V_2 = 0$, $V_1 = 0$ on the nodal voltage equations

$$Y_{11}V_1+Y_{12}V_2 = I_1$$
$$Y_{21}V_1+Y_{22}V_2 = I_2$$

When $V_2 = 0$, which corresponds to the short-circuiting of terminals d and s,

$$I_1 = s(C_{gs}+C_{gd})V_1$$
$$I_2 = g_mV_1-sC_{gd}V_1$$

whence

$$Y_{11} = I_1/V_1|_{V_t = 0} = s(C_{gs}+C_{gd})$$
$$Y_{21} = I_2/V_1|_{V_2 = 0} = g_m-sC_{gd}$$

Similarly, putting $V_1 = 0$ (which causes the voltage-controlled current-source g_mV_1 to vanish),

$$Y_{12} = I_1/V_2|_{V_1 = 0} = -sC_{gd}$$
$$Y_{22} = I_2/V_2|_{V_1 = 0} = g_d+s(C_{ds}+C_{gd})$$

or

$$[Y]_{cs}= \begin{bmatrix} s(C_{gs} + C_{gd}) & -sC_{gd} \\ g_m-sC_{gd} & g_d+s(C_{ds}+C_{gd}) \end{bmatrix}$$

Finding the matrix elements in this way from first principles involves the nodal-voltage equations for the active network as a whole. It is therefore interesting to consider an alternative approach that renders the matrix obvious.

Figure 1.25 can be represented as two networks in parallel. One comprises the purely passive, reciprocal elements, and the other the pure voltage-controlled current source g_mV_1, as shown in Figures 1.26(a) and (b).

Figure 1.26(a) has the topology of a π network, and as it is passive, by inspection its matrix is

$$[Y]_a = \begin{bmatrix} s(C_{gs}+C_{gd}) & -sC_{gd} \\ -sC_{gd} & g_d+s(C_{ds}+C_{gd}) \end{bmatrix}$$

Figure 1.26(b) is entirely unilateral. $I_1 = 0$, and when $V_1 = 0$, $g_mV_1 = I_2 = 0$. Hence the matrix is simply

$$[Y]_b = \begin{bmatrix} 0 & 0 \\ g_m & 0 \end{bmatrix}$$

and the addition of $[Y]_b$ to $[Y]_a$ gives the same result as before. (b) For a load Y_2, the second nodal-voltage equation becomes

$$Y_{21}V_1+Y_{22}V_2 = I_2 = -Y_2V_2$$

whence

$$\frac{V_2}{V_1} = -\frac{Y_{21}}{Y_{22}+Y_2}$$

Substituting values then gives

$$\frac{V_2}{V_1} = -\frac{(2-j0{\cdot}1)10^{-3}}{(1{\cdot}02+j1)10^{-4}} = -14 \underline{/-47{\cdot}3°}$$

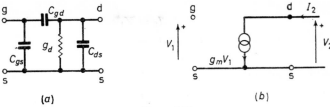

Figure 1.26

Comment

(1) The formulation of the Y-matrix on a 2-port basis should be compared with the approach in Illustration 1.11. The matrix is essentially an organised set of admittance coefficients, and with the load excluded, the result is the same either way. However, the two-port approach has the advantage of generality, for the network matrix is defined solely by voltage constraints (in this case) that are independent of any particular source or load that may be used in practice. Other matrix forms are similarly defined by constraints on currents, as for the Z-matrix; or by both current and voltage constraints, as, for example, in the case of the hybrid or h-matrix.

(2) A matrix displays the pattern of the coefficients in an ordered set of simultaneous equations. The second procedure illustrated may be likened to 'stamping' the obvious pattern for the unilateral voltage-controlled current source ($g_m V_1$) onto the obvious pattern for the passive π-network: the components of the patterns are self-aligning.

ILLUSTRATION 1.17

The external behaviour of a linear two-port network is expressed by the matrix equation,

$$\begin{bmatrix} h_{11} & h_{12} \\ h_{21} & h_{22} \end{bmatrix} \cdot \begin{bmatrix} I_1 \\ V_2 \end{bmatrix} = \begin{bmatrix} V_1 \\ I_2 \end{bmatrix}$$

Transform the h-matrix into a Z-matrix.

Interpretation

The corresponding equations are

$$h_{11}I_1 + h_{12}V_2 = V_1 \tag{1a}$$

$$h_{21}I_1 + h_{22}V_2 = I_2 \tag{1b}$$

and

$$Z_{11}I_1 + Z_{12}I_2 = V_1 \qquad (2a)$$
$$Z_{21}I_1 + Z_{22}I_2 = V_2 \qquad (2b)$$

Transposing equation (1b) gives

$$(-h_{21}/h_{22})I_1 + (1/h_{22})I_2 = V_2 \qquad (3)$$

whence, by comparing coefficients with equation (2b),

$$Z_{21} = -h_{21}/h_{22} \quad \text{and} \quad Z_{22} = 1/h_{22}$$

Then substituting for V_2 from equation (3) into equation (1a) and comparing coefficients with equation (2a) gives

$$Z_{11} = h_{11} - (h_{12}h_{21})/h_{22} \quad \text{and} \quad Z_{12} = h_{12}/h_{22} \qquad (4)$$

Comment

The h or hybrid parameters, so termed because of their mixed dimensions, are often thought to be peculiar to transistors, to which they are well suited for the specification of small signal (approximately linear) behaviour. They were conceived, however, long before transistors, and equations (1) exemplify yet another of the six possible ways of inter-relating the input and output variables for any linear two-port network.

Each kind of two-port matrix has a unique role in relation to groups of two-port networks (see Illustration 1.15), and it is important to be able to transform from one kind into another. This may be facilitated with a matrix conversion chart; but often, as shown here, it is easy to make the transformation directly.

ILLUSTRATION 1.18

A transistor (type BC187), when operated with the collector at -5 V relative to the emitter and biassed for a collector current of 2 mA, has low-frequency h parameters $h_{ie} = 2$ kΩ, $h_{re} = 1 \cdot 4 \times 10^{-4}$, $h_{fe} = 140$ and $h_{oe} = 29$ μS. It is used as a common-emitter amplifier for low frequency alternating voltages of the order of a few millivolts, with biassing circuitry and a dc supply arranged to maintain the stated operating conditions with a collector load resistance $R_c = 4$ kΩ and an un-bypassed resistor $R_e = 500$ Ω in series with the emitter. Find the voltage gain and input impedance.

Interpretation

Solution with h-parameters is awkward; but the arrangement may be viewed as a simple case of series-connected two-port networks by the expedient of treating R_e as a rudimentary two-port network as indicated in Figure 1.27. This arrangement is subject to easy solution by Z-matrix addition, and it is simple to convert the given transistor h-parameters into Z-parameters.

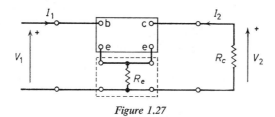

Figure 1.27

Let $[Z]_e$ and $[Z]_T$ denote the matrices for the two-port network artificially constituted of R_e, and the transistor, respectively. Then by inspection,

$$[Z]_e = \begin{bmatrix} R_e & R_e \\ R_e & R_e \end{bmatrix} = \begin{bmatrix} 500 & 500 \\ 500 & 500 \end{bmatrix}$$

and using the formulae derived in Illustration 1.17, noting that h_{ie}, h_{re}, h_{fe} and h_{oe} are respectively equivalent to h_{11}, h_{12}, h_{21} and h_{22},

$$[Z]_T = \begin{bmatrix} 1325 & 4\cdot83 \\ -4\cdot83\times10^6 & 3\cdot45\times10^4 \end{bmatrix}$$

The total matrix for Figure 1.27 is thus

$$[Z] = [Z]_e + [Z]_T = \begin{bmatrix} 1825 & 505 \\ -4\cdot83\times10^6 & 3\cdot50\times10^4 \end{bmatrix}$$

The equations to be solved have the compact matrix for $[Z]\cdot[I] = [V]$ or the algebraic forms

$$Z_{11}I_1 + Z_{12}I_2 = V_1$$
$$Z_{21}I_1 + Z_{22}I_2 = V_2 = -R_cI_2$$

and the arithmetical forms

$$1825I_1 + 505I_2 = V_1$$
$$-483I_1 + 3\cdot90I_2 = 0$$

whence

$$A = \frac{V_2}{V_1} = \frac{-R_c I_2}{V_1} = -7\cdot71$$

$$Z_{in} = \frac{V_1}{I_1} = 64\cdot4 \text{ k}\Omega$$

Comment

(1) The incorporation of R_e has been approached formally; but it is almost obvious that it must contribute equally to the self and transfer elements of the transistor mesh-type impedance matrix. Note that this case is the dual of a parallel-connected admittance as in Illustration 6.6.

(2) The effect of R_e is most marked on Z_{11} and R_{12}. A transistor exhibits a degree of reverse transmission, or instrinsic feedback, compatible with non-zero values for h_{re} (or h_{12}), Z_{12} or Y_{12}. The addition of R_e increases this feedback, and in this particular circuit its effect is to diminish the gain but increase the input impedance. This is consistent with current dependent series-applied negative voltage feedback (see Illustration 6.1). If R_e is by-passed with a capacitor of negligible reactance so that $Z_e \simeq 0$, the gain rises to -260 while the input impedance falls to $1930\ \Omega$.

ILLUSTRATION 1.19

A transistor (type BC 187), when operated with the collector at -5 V relative to the emitter and biassed for a collector current of 2 mA, has low-frequency h-parameters $h_{ie} = 2$ kΩ, $h_{re} = 1\cdot4\times10^{-4}$, $h_{fe} = 140$ and $h_{oe} = 29$ μS.

(1) Determine the indefinite admittance matrix $[Y]_i$ for the stated operating conditions.

(2) Find the Y-parameters for the transistor when operated in the common-collector configuration, and calculate the output impedance between emitter and collector when the base is connected to the collector through a signal source of resistance 10 kΩ.

Interpretation

(1) The h-parameters are first transformed into Y-parameters using the relations

$$Y_{ie} = 1/h_{ie} = 5\times10^{-4} \text{ S}$$
$$Y_{re} = -h_{re}/h_{ie} = -7\times10^{-8} \text{ S}$$
$$Y_{fe} = h_{fe}/h_{ie} = 7\times10^{-2} \text{ S}$$
$$Y_{oe} = h_{oe}-(h_{fe}h_{re}/h_{ie}) = 1\cdot92\times10^{-5} \text{ S}$$

(see Illustration 6.6).

Let the transistor electrodes b, c, e be identified respectively with nodes numbered 1, 2, 3, as in Figure 1.4(b). Then the matrix of common-emitter Y-parameters is the matrix for node 3 as datum, and corresponds to the deletion of the row and column containing Y_{33} from Y_i; or,

$$[Y]_{ce} = \begin{bmatrix} 5 \times 10^{-4} & -7 \times 10^{-8} \\ 7 \times 10^{-2} & 1.92 \times 10^{-5} \end{bmatrix} = \begin{bmatrix} Y_{11} & Y_{12} & Y_{13} \\ Y_{21} & Y_{22} & Y_{23} \\ Y_{31} & Y_{32} & Y_{33} \end{bmatrix} \quad (1)$$

The zero-sum property of $[Y]_i$ is obeyed by non-reciprocal networks provided they are linear, in the sense that inequalities of the form $Y_{ik} \neq Y_{ki}$ are attributable to the existance of an excitation-controlled source, such as a voltage-controlled current source, in a circuit-model of otherwise linear immittances. Thus, to three significant figures, the deleted elements are

$$Y_{31} = -(Y_{11}+Y_{21}) = -7.05 \times 10^{-2}$$
$$Y_{32} = -(Y_{12}+Y_{22}) = -1.91 \times 10^{-5}$$
$$Y_{23} = -(Y_{21}+Y_{22}) = -7 \times 10^{-2}$$
$$Y_{13} = -(Y_{11}+Y_{12}) = -5 \times 10^{-4}$$
$$Y_{33} = -(Y_{13}+Y_{23}) = -(Y_{31}+Y_{32}) = 7.05 \times 10^{-2}$$

Therefore,

$$[Y]_i = \begin{bmatrix} 5 \times 10^{-4} & -7 \times 10^{-8} & -5 \times 10^{-4} \\ 7 \times 10^{-2} & 1.92 \times 10^{-5} & -7 \times 10^{-2} \\ -7.05 \times 10^{-2} & -1.91 \times 10^{-5} & 7.05 \times 10^{-2} \end{bmatrix} \quad (2)$$

(2) The matrix of Y-parameters for the transistor in the common-collector configuration is the matrix for electrode c or node 2 as datum, and is given by deleting from Y_i the row and column containing the element Y_{22}. Thus,

$$[Y]_{cc} = \begin{bmatrix} 5 \times 10^{-4} & -5 \times 10^{-4} \\ -7.05 \times 10^{-2} & 7.05 \times 10^{-2} \end{bmatrix} \quad (3)$$

The output impedance of the common-collector circuit may now be found by applying a voltage V_{ec} between emitter and collector and calculating its response current I_e while the base-collector signal-source emf is suppressed. Let V_{bc} denote the voltage across the internal resistance (10 kΩ) of the signal-source. Then, using the usual two-port polarity convention (b^+, e^+), with the base current I_b in the sense of V_{bc},

$$5\times10^{-4}V_{bc} - 5\times10^{-4}V_{ec} = I_b = -V_{bc}\times10^{-4}$$
$$-7{\cdot}05\times10^{-2}V_{bc}+7{\cdot}05\times10^{-2}V_{ec} = I_e$$

whence

$$Z_\text{out} = V_{ec}/I_e = 85{\cdot}1\ \Omega$$

Comment

This illustration demonstrates the great importance of the indefinite admittance matrix as a simple, systematic means for determining the behaviour of a multi-port network when any pair of its ports is chosen as an input and an output port.

Note that the slight departure from the zero-sum property in equation (2) is due only to working to a limited but adequate number of significant figures.

The validity of the zero-sum property in a non-reciprocal case such as this is easily confirmed by reference to a simple equivalent representation for a device that is non-reciprocal but linear in the range of operation considered. Such a representation might be a three-node network of linear admittances Y_a, Y_b and Y_c linking nodes 1 and 2, 1 and 3, 2 and 3, respectively, with a voltage-controlled current source $g_m(V_1-V_3)$ in parallel with Y_c and flowing from node 2 to node 3. Then, with reference to an external datum,

$$\begin{bmatrix} Y_b+Y_b & -Y_a & -Y_b \\ -Y_a & Y_a+Y_c & -Y_c \\ -Y_b & -Y_c & Y_b+Y_c \end{bmatrix} \cdot \begin{bmatrix} V_1 \\ V_2 \\ V_3 \end{bmatrix} = \begin{bmatrix} I_1 \\ I_2-g_m(V_1-V_3) \\ I_3+g_m(V_1-V_3) \end{bmatrix}$$

or more compactly, writing $Y_{11} = Y_a+Y_b$ etc.,

$$\begin{bmatrix} Y_{11} & Y_{12} & Y_{13} \\ Y_{21}+g_m & Y_{22} & Y_{23}-g_m \\ Y_{31}-g_m & Y_{32} & Y_{33}+g_m \end{bmatrix} \cdot \begin{bmatrix} V_1 \\ V_2 \\ V_3 \end{bmatrix} = \begin{bmatrix} I_1 \\ I_2 \\ I_3 \end{bmatrix}$$

It is evident that g_m cancels in summing along any row or column, and the remaining elements are reciprocal and known to possess the zero-sum property.

ILLUSTRATION 1.20

The external behaviour of a linear two-port network may be expressed either in the form $[Z]\cdot[I] = [V]$ or in the form $[Y]\cdot[V] = [I]$. Find the elements of each immittance matrix in terms of the other (a) by imposing suitable constraints on the voltages or currents; and(b), by solving for the currents or or voltages with Cramer's rule applied to the determinant of the matrix.

Interpretation

The equations corresponding to the given matrix forms are

$$Z_{11}I_1 + Z_{12}I_2 = V_1$$
$$Z_{21}I_1 + Z_{22}I_2 = V_2 \tag{1}$$

and

$$Y_{11}V_1 + Y_{12}V_2 = I_1$$
$$Y_{21}V_1 + Y_{22}V_2 = I_2 \tag{2}$$

(a) Imposing voltage constraints simultaneously on both pairs of equations,

$$Y_{11} = \frac{I_1}{V_1}\bigg|_{V_2 = 0} = \frac{V_1 Z_{22}}{D} \cdot \frac{1}{V_1} = \frac{Z_{22}}{D}$$

where $D = Z_{11}Z_{22} - Z_{21}Z_{12}$

$$Y_{21} = \frac{I_2}{V_1}\bigg|_{V_2 = 0} = \frac{V_1 Z_{21}}{-D} \cdot \frac{1}{V_1} = \frac{-Z_{21}}{D}$$

$$Y_{12} = \frac{I_1}{V_2}\bigg|_{V_1 = 0} = \frac{V_2 Z_{12}}{-D} \cdot \frac{1}{V_2} = \frac{-Z_{12}}{D}$$

$$Y_{22} = \frac{I_2}{V_2}\bigg|_{V_1 = 0} = \frac{V_2 Z_{11}}{D} \cdot \frac{1}{V_2} = \frac{Z_{11}}{D}$$

The common denominator $D = Z_{11}Z_{22} - Z_{21}Z_{12}$ is recognisable as the determinant Δ_z of $[Z]$, so that $[Y]$ may be written compactly as

$$[Y] = \begin{bmatrix} Y_{11} & Y_{12} \\ Y_{21} & Y_{22} \end{bmatrix} = \frac{1}{\Delta_z} \begin{bmatrix} Z_{22} & -Z_{12} \\ -Z_{21} & Z_{11} \end{bmatrix} \tag{3}$$

Since equations (1) and (2) are duals, the reverse transformation is given immediately on exchanging symbols in equation (3). Thus,

$$[Z] = \begin{bmatrix} Z_{11} & Z_{12} \\ Z_{21} & Z_{22} \end{bmatrix} = \frac{1}{\Delta_y} \begin{bmatrix} Y_{22} & -Y_{12} \\ -Y_{21} & Y_{11} \end{bmatrix} \tag{4}$$

where $\Delta_y = \det[Y] = Y_{11}Y_{22} - Y_{21}Y_{12}$

(b) Applying Cramer's rule to equations (1), for example,

$$I_1 = \frac{\Delta_{11}}{\Delta_z} V_1 + \frac{\Delta_{21}}{\Delta_z} V_2 = Y_{11}V_1 + Y_{12}V_2$$

$$I_2 = \frac{\Delta_{12}}{\Delta_z} V_1 + \frac{\Delta_{22}}{\Delta_z} V_2 = Y_{21}V_1 + Y_{22}V_2$$

where the cofactors of Δ_z are

$$\Delta_{11} = Z_{22}, \quad \Delta_{21} = -Z_{12}, \quad \Delta_{12} = -Z_{21} \quad \text{and} \quad \Delta_{22} = Z_{11}$$

Thus,

$$Y = \frac{1}{\Delta_z} \begin{bmatrix} \Delta_{11} & \Delta_{21} \\ \Delta_{12} & \Delta_{22} \end{bmatrix} = \frac{1}{\Delta_z} \begin{bmatrix} Z_{22} & -Z_{12} \\ -Z_{21} & Z_{11} \end{bmatrix}$$

as before. The solution of eqns. (2) by Cramer's rule leads similarly to eqn. (4).

Comment

These transformations are examples of matrix inversion. As a matrix is essentially a pattern, it is not subject to the commutative laws of normal algebra. Thus, while a family of equations might be written in the compact form

$$[Z] \cdot [I] = [V]$$

transposition in a symbolic sense only is valid, in the form

$$[I] = [Z]^{-1} \cdot [V] = [Y] \cdot [V]$$

where $[Y] = [Z]^{-1}$ is not $1/Z$ but is a new array of elements, related to the original and called its inverse.

The general form for the inverse of a square matrix of order n is given by applying Cramer's rule to n simultaneous equations. Given

$$[V] = [Z] \cdot [I]$$

or, expanded,

$$\begin{bmatrix} V_1 \\ V_2 \\ \cdot \\ V_n \end{bmatrix} = \begin{bmatrix} Z_{11} & Z_{12} \ldots Z_{1n} \\ Z_{21} & Z_{22} \ldots Z_{2n} \\ \cdot & \cdot\cdot\; \cdots\;\; \cdot \\ Z_{n1} & Z_{n2} \ldots Z_{nn} \end{bmatrix} \cdot \begin{bmatrix} I_1 \\ I_2 \\ \cdot\cdot \\ I_n \end{bmatrix}$$

then by Cramer's rule,

$$\begin{bmatrix} I_1 \\ I_2 \\ . \\ I_n \end{bmatrix} = \frac{1}{\Delta_z} \cdot \begin{bmatrix} \Delta_{11} & \Delta_{21} \ldots \Delta_{n1} \\ \Delta_{12} & \Delta_{22} \ldots \Delta_{n2} \\ . & \ldots \\ \Delta_{1n} & \Delta_{2n} \ldots \Delta_{nn} \end{bmatrix} \cdot \begin{bmatrix} V_1 \\ V_2 \\ . \\ V_n \end{bmatrix}$$

or, compressed,

$$[I] = [Z]^{-1} \cdot [V] = [Y] \cdot [V]$$

The inverse of a square matrix such as $[Z]$ is more generally expressed by

$$[Z]^{-1} = \frac{\text{Adj}[Z]}{\Delta_z}$$

where $\text{Adj}[Z]$ is the *adjoint* of $[Z]$, formed by replacing each element in $[Z]$ by the cofactor of that element, and then by exchanging rows with columns in the resultant array to form a *transpose matrix*. This method of inversion is not practicable, however, for high order matrices (eg $n > 3$) owing to the rate in growth of cofactors with n.

ILLUSTRATION 1.21

Two linear systems are governed respectively by equations of the forms

$$\begin{aligned} p_1 &= a_{11}p_2 + a_{12}q_2 \\ q_1 &= a_{21}p_2 + a_{22}q_2 \end{aligned} \quad \text{or} \quad \begin{bmatrix} p_1 \\ q_1 \end{bmatrix} = [A] \cdot \begin{bmatrix} p_2 \\ q_2 \end{bmatrix} \tag{1}$$

and

$$\begin{aligned} p_3 &= b_{11}p_4 + b_{12}q_4 \\ q_3 &= b_{21}p_4 + b_{22}q_4 \end{aligned} \quad \text{or} \quad \begin{bmatrix} p_3 \\ q_3 \end{bmatrix} = [B] \cdot \begin{bmatrix} p_4 \\ q_4 \end{bmatrix} \tag{2}$$

where p_1, q_1 and p_3, q_3 are input or excitation variables while p_2, q_2 and p_4, q_4 are output or response variables.

The systems are connected in sequence so that the output variables p_2, q_2 from the first provide the excitation variables p_3, q_3 for the second. Determine the matrix of coefficients for the composite system, and thence attribute meaning to the concept of matrix multiplication.

Interpretation

Putting $p_2 = p_3$ and $q_2 = q_3$ in the matrix forms of equations (1) and (2) and substituting from (2) into (1) gives

$$\begin{bmatrix} p_1 \\ p_2 \end{bmatrix} = [A] \cdot \begin{bmatrix} p_3 \\ q_3 \end{bmatrix} = [A] \cdot [B] \cdot \begin{bmatrix} p_4 \\ q_4 \end{bmatrix} = [C] \cdot \begin{bmatrix} p_4 \\ q_4 \end{bmatrix} \qquad (3)$$

A matrix, being an array of coefficients or variables, cannot possess numerical value as an entity; and therefore a matrix product such as $[A] \cdot [B]$ cannot be likened to the product of single algebraic symbols to which numerical values can be assigned. However, the special significance of the symbolic product $[A] \cdot [B]$ may be seen by reverting to equations (1) and (2) in their direct forms. Making the same substitutions as before in these equations gives

$$p_1 = (a_{11}b_{11} + a_{12}b_{21})p_4 + (a_{11}b_{12} + a_{12}b_{22})q_4$$
$$q_1 = (a_{21}b_{11} + a_{22}b_{21})p_4 + (a_{21}b_{12} + a_{22}b_{22})q_4$$

or in compressed matrix notation,

$$\begin{bmatrix} p_1 \\ q_1 \end{bmatrix} = \begin{bmatrix} c_{11} & c_{12} \\ c_{21} & c_{22} \end{bmatrix} \cdot \begin{bmatrix} p_4 \\ q_4 \end{bmatrix} = [C] \cdot \begin{bmatrix} p_4 \\ q_4 \end{bmatrix} = [A] \cdot [B] \cdot \begin{bmatrix} p_4 \\ q_4 \end{bmatrix}$$

Thus, the symbolic product $[A] \cdot [B]$ has meaning in the systematic row-column multiplication represented by

$$[C] = \begin{bmatrix} c_{11} & c_{12} \\ c_{21} & c_{22} \end{bmatrix} = [A] \cdot [B]$$

$$= \begin{bmatrix} a_{11} & a_{12} \\ a_{21} & a_{22} \end{bmatrix} \cdot \begin{bmatrix} b_{11} & b_{12} \\ b_{21} & b_{22} \end{bmatrix}$$

$$= \begin{bmatrix} a_{11}b_{11} + a_{12}b_{21} & a_{11}b_{12} + a_{12}b_{22} \\ a_{21}b_{11} + a_{22}b_{21} & a_{21}b_{12} + a_{22}b_{22} \end{bmatrix}$$

Comment

The product $[A] \cdot [B]$ representing the pattern of coefficients for the composite system in terms of the coefficients for its parts is non-commutative, or $[A] \cdot [B] \neq [B] \cdot [A]$. This accords with the proposition that if the system 1 differs from system 2, the composite behaviour when 2 follows 1 may be expected to differ from that exhibited when 2 precedes 1. The product of matrices is in general non-commutative. In this illustration $[B]$ may be said to post-multiply $[A]$, or $[A]$ to pre-multiply $[B]$.

Matrices to be multiplied together must be conformable, in the sense that the second or post-multiplying one must have the same number of rows as the first has columns.

ILLUSTRATION 1.22

The equations for a two-port network corresponding to equations (1) in Illustration 1.21 have the forms

$$V_1 = a_{11}V_2 - a_{12}I_2$$
$$I_1 = a_{21}V_2 - a_{22}I_2$$
$$\text{or} \quad \begin{bmatrix} V_1 \\ I_1 \end{bmatrix} = \begin{bmatrix} a_{11} & a_{12} \\ a_{21} & a_{22} \end{bmatrix} \cdot \begin{bmatrix} V_2 \\ -I_2 \end{bmatrix}$$

for the usual two-port polarity convention.

(1) Show that the a-matrices for Figures 1.28(a) and 1.28(b) are

$$[a]_Z = \begin{bmatrix} 1 & Z \\ 0 & 1 \end{bmatrix} \quad \text{and} \quad [a]_Y = \begin{bmatrix} 1 & 0 \\ Y & 1 \end{bmatrix}$$

Thence obtain the a-matrices for a T-network, of branches Z_1 (series left), Y (shunt) and Z_2 (series right); and for a π-network, of branches Y_1 (shunt left), Z (series) and Y_2 (shunt right).

(2) In Figure 1.28(c) the network N, which is passive, reciprocal and balanced-to-ground, is specified by

$$[a]_N = \begin{bmatrix} -3 & -20 \\ -0.4 & -3 \end{bmatrix}$$

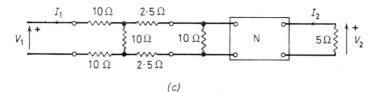

(a)   (b)

(c)

Figure 1.28

Find V_2/V_1, and I_2/V_1 when the load on N is replaced by a short-circuit.

(3) A CE low-frequency transistor amplifier comprises two stages. Each, including its collector load, is defined by

$$[a]_{CE} = \begin{bmatrix} -66{\cdot}5\times10^{-4} & -30{\cdot}6 \\ -5{\cdot}1\times10^{-6} & -2{\cdot}04\times10^{-2} \end{bmatrix}$$

Find the voltage gain and input impedance of the amplifier.

Interpretation

(1) For Figures 1.28(a) and (b), the equations $V_1 = a_{11}V_2 - a_{12}I_2$, $I_1 = a_{21}V_2 - a_{22}I_2$ become

$$V_1 = V_2 - ZI_2 \qquad \text{or} \qquad \begin{bmatrix} V_1 \\ I_1 \end{bmatrix} = \begin{bmatrix} 1 & Z \\ 0 & 1 \end{bmatrix} \cdot \begin{bmatrix} V_2 \\ -I_2 \end{bmatrix}$$
$$I_1 = 0 - I_2$$

and

$$V_1 = V_2 - 0 \qquad \text{or} \qquad \begin{bmatrix} V_1 \\ I_1 \end{bmatrix} = \begin{bmatrix} 1 & 0 \\ Y & 1 \end{bmatrix} \cdot \begin{bmatrix} V_2 \\ -I_2 \end{bmatrix}$$
$$I_1 = YV_2 - I_2$$

Hence

$$[a]_Z = \begin{bmatrix} 1 & Z \\ 0 & 1 \end{bmatrix} \quad \text{and} \quad [a]_Y = \begin{bmatrix} 1 & 0 \\ Y & 1 \end{bmatrix} \tag{1}$$

The T and π-networks may be synthesised from Figures 1.28(a) and (b). Using the matrix multiplication procedure evolved in Illustration 1.21,

$$\begin{aligned} [a]_T &= \begin{bmatrix} 1 & Z_1 \\ 0 & 1 \end{bmatrix} \cdot \begin{bmatrix} 1 & 0 \\ Y & 1 \end{bmatrix} \cdot \begin{bmatrix} 1 & Z_2 \\ 0 & 1 \end{bmatrix} \\ &= \begin{bmatrix} 1+YZ_1 & Z_1 \\ Y & 1 \end{bmatrix} \cdot \begin{bmatrix} 1 & Z_2 \\ 0 & 1 \end{bmatrix} \\ &= \begin{bmatrix} 1+YZ_1 & Z_1+Z_2+YZ_1Z_2 \\ Y & 1+YZ_2 \end{bmatrix} \end{aligned} \tag{2}$$

Similarly,

$$\begin{aligned} [a]_\pi &= \begin{bmatrix} 1 & 0 \\ Y_1 & 1 \end{bmatrix} \cdot \begin{bmatrix} 1 & Z \\ 0 & 1 \end{bmatrix} \cdot \begin{bmatrix} 1 & 0 \\ Y_2 & 1 \end{bmatrix} \\ &= \begin{bmatrix} 1+Y_2Z & Z \\ Y_1+Y_2+ZY_1Y_2 & 1+Y_1Z \end{bmatrix} \end{aligned} \tag{3}$$

(2) The network preceding N is the balanced H-form of a T-network, necessary for association with N in a system having no common or earth line. As it is formed from a T-network by apportioning the series arms equally between top and bottom, its matrices are the same as those for the corresponding T-network, and by equation (2),

$$[a]_H = [a]_T = \begin{bmatrix} 3 & 35 \\ 0\cdot1 & 1\cdot5 \end{bmatrix}$$

Then for the whole system,

$$[a] = [a]_H \cdot [a]_Y \cdot [a]_N$$

$$= \begin{bmatrix} 3 & 35 \\ 0\cdot1 & 1\cdot5 \end{bmatrix} \cdot \begin{bmatrix} 1 & 0 \\ 0\cdot1 & 1 \end{bmatrix} \cdot \begin{bmatrix} -3 & -20 \\ -0\cdot4 & -3 \end{bmatrix} = \begin{bmatrix} -33\cdot5 & -235 \\ -1\cdot35 & -9\cdot5 \end{bmatrix}$$

and

$$\begin{bmatrix} V_1 \\ I_1 \end{bmatrix} = \begin{bmatrix} -33\cdot5 & -235 \\ -1\cdot35 & -9\cdot5 \end{bmatrix} \cdot \begin{bmatrix} V_2 \\ -I_2 \end{bmatrix}$$

or

$$V_1 = -33\cdot5 V_2 + 235 I_2$$
$$I_1 = -1\cdot35 V_2 + 9\cdot5 I_2$$

The load is incorporated by writing $V_2 = -5I_2$, or $I_2 = -0\cdot2V_2$ (V_2 is a potential fall due to I_2, not a rise as assumed in setting-up the equations), giving

$$V_2/V_1 = -1/80\cdot5$$

When the output terminals are short-circuited, $V_2 = 0$, and

$$\left. \frac{I_2}{V_1} \right|_{V_2 = 0} = Y_{21} = \frac{1}{235}$$

(3) The matrix for the whole amplifier has the form

$$[a] = [a]_{CE} \cdot [a]_{CE} = \{[a]_{CE}\}^2$$

Multiplying the given matrix of numerical elements by itself once gives

$$[a] = \begin{bmatrix} 2\times10^{-4} & 82\cdot5\times10^{-2} \\ 13\cdot8\times10^{-8} & 5\cdot72\times10^{-4} \end{bmatrix}$$

Since the load has been included in the matrix, $I_2 = 0$ (externally). Therefore,

$$\frac{V_2}{V_1} = \frac{1}{a_{11}} = \frac{10^4}{2} = 5\times10^3$$

and $I_1 = a_{21}V_2 = a_{21}\times5\times10^3 V_1$, or

$$Z_{in} = \frac{V_1}{I_1} = \frac{1}{13\cdot8\times10^{-8}\times5\times10^3} = 1\cdot45 \text{ k}\Omega$$

Comment

The ordered elements $a_{11} \ldots a_{22}$ are equivalent to the long-established $A\,B\,C\,D$ parameters, and convenience may dictate choice. Note that the sign of each element is independent of polarity convention: this is the reason for attaching a negative sign to I_2 when shown entering the network, as in the customary two-port convention. By contrast, the signs of Z_{12}, Z_{21} or Y_{12}, Y_{21} depend on convention.

Observe that for the linear passive, reciprocal networks in cases (1) and (2), det $[a] = a_{11}a_{22} - a_{21}a_{12} = AD - BC = 1$ This property is general for such networks: see Illustration 3.17. It does not apply to non-reciprocal devices, which is apparant in the case of the transistor amplifier. See also Illustration 6.10.

CHAPTER 2

Transient Response and its correlation with Frequency Response

INTRODUCTION

The objective of this chapter is to illustrate, with simple passive and active networks, the significant features of response as a function of time, or response in the so-called time domain. The most important feature is the natural behaviour and its natural modes or frequencies; for as these are intrinsic, they define the network or system independently of external excitations. Thus, while the natural behaviour is in losing conflict with the force of any imposed continuous excitation, it governs response at the inception of the transient period, when such an excitation is abruptly applied, and is fundamental to the concepts of growth and decay. The natural behaviour is therefore inherent in considerations of stability, and in the case of active networks provides a simple, basic criterion of regeneration (see Illustration 2.10).

The natural frequencies of a network or system are also the links between response in the time and frequency domains. They are identifiable with the poles and zeros of impedance and admittance functions, and through them alone it is possible, without involvement of the actual circuit elements, to determine the response to a sinusoidal excitation in the steady state (see Illustrations 2.8 and 2.11). In reverse, this is the basis of network synthesis.

Correlation between time and frequency response is shown equivalently by reference to the Fourier transform, particularised for a step function, in Illustration 2.7; and by the Laplace transform, in

Illustrations 2.12, 2.13 and 2.14. These transforms are not considered in detail, for the motive is simply to highlight their roles. The Laplace transform, in rendering differential equations algebraic, has often been regarded as a marked simplification; but this is not always true. It may lead to algebraic complexity that must be weighed against the quotability of solution-forms for linear homogeneous differential equations, in which often the unknown constants may be easy to find from initial conditions (see Illustrations 2.9, 2.10 and 2.14). Moreover, the digital computer is well adapted to the complete solution of a family of first-order differential equations in the time (or other) domain, and interest has reverted to differential equations and to the resolution of the nth order equation for a complex system into n first order ones expressed in matrix form. This is the essence of the modern *state-variable* technique.

In the following illustrations, a notation of form $F(0)$, such as $i(0)$, $v(0)$, $q(0)$, implies an initial value at an instant so infinitesimally removed on the positive side from the boundary $t = 0$ as to be coincident with the inception of an excitation at $t = 0$. This initial value may originate from the value of a quantity at the moment it impinges from the negative side ($t < 0$), on to the boundary $t = 0$. The notation $F(0)$ may be taken as equivalent to $F(0+)$, as often used in this context.

ILLUSTRATION 2.1

A unidirectional and constant emf E (as from an ideal battery) is applied to an inductor L in series with a resistor R, at the instant designated for convenience as $t = 0$. The resistance of the inductor, which would be inseparable from L in practice, may be regarded as included in R.

(a) Find the natural mode for the circuit, the transient and steady-state components of the current, and the actual current at any instant $t > 0$.

(b) E vanishes at an instant $t = t_1$, but the circuit remains closed. Find an expression for the current at $t > t_1$.

(c) Find expressions for the voltage across the inductor under conditions (a) and (b) that would be valid if its self-resistance were negligible.

Interpretation

(a) The abrupt excitation of a circuit or system may be symbolised with a unit step-function, $U(t)$. This is zero for $t < 0$ and unity for $t > 0$. In general the symbolism has the form $U(t) F(t)$, where $F(t)$

denotes a time-varying excitation. In the present case, application of the emf E at $t = 0$ is symbolised by $U(t)E$, and the KVL equation is formally

$$L \frac{di(t)}{dt} + Ri(t) = U(t)E \qquad (1)$$

This may be written more concisely as

$$L\,Di + Ri = U(t)E \qquad (2)$$

where D is the differential operator d/dt and $i(t)$ is abbreviated to i.

$$U(t)E = 0, \quad t < 0 \quad \text{but} \quad U(t)E = E, \quad t > 0.$$

The conditions immediately after the application of E and subsequently are therefore defined by

$$L\,Di + Ri = E \qquad (t > 0) \qquad (3)$$

The *natural* or *excitation-free* behaviour of the circuit (as distinct from that ultimately forced on it by the excitation) is governed by the *homogeneous* equation

$$L\,Di + Ri = 0 \qquad (4)$$

An nth order homogeneous equation for a linear system is satisfied by a solution comprising n exponential terms of the form $A_n e^{s_n t} + A_{n-1} e^{s_{n-1} t} + \ldots + A_1 e^{s_1 t}$. In this first order case let $i = Ae^{st}$. Equation (4) then becomes

$$(Ls + R)Ae^{st} = 0 \qquad (5)$$

which can be true for i finite only if

$$Ls + R = 0 \qquad (6)$$

This *characteristic equation* has a single root,

$$s = -R/L \qquad (7)$$

which is the *natural mode* or *natural frequency*.

The transient component i_{tr} of the complete solution in response to the excitation E is synonymous in form with natural behaviour, or

$$i_{tr} = Ae^{st} = Ae^{-RT/L} \qquad (8)$$

where A is determined by initial conditions set wholly or partly, according to the passivity or activity of the elements themselves at the moment $t = 0$, by the initial magnitude of the excitation (in this case E).

The steady-state component i_{ss} is synonymous with the *particular integral*, derived from

$$L\,Di_{ss} + Ri_{ss} = E \qquad (9)$$

Equation (9) may be transposed in its operational form to give

$$i_{ss} = \frac{E}{L} \cdot \frac{1}{D-\alpha} = \frac{E}{R} \tag{10}$$

where $\alpha = R/L$ and the standard result for the operation $1/(D-\alpha)$ on a constant has been invoked. Such formality is, however, pedantic; for i_{ss} is obviously E/R: in a closed circuit with a time-invariant emf. the current when steady must be independent of the series inductance; for this produces a counter-potential limiting the current only so long as it is changing, with an instantaneous value $i = (E-L\,Di)/R$.

The actual current at any instant t is

$$i = i_{tr} + i_{ss} \tag{11}$$

Let the circuit elements be initially passive. For circuits in general, this means that at the moment of excitation ($t = 0$) inductors are devoid of the flux-linkages ($\Lambda = Li = 0$) and the capacitors are uncharged ($q = Cv = 0$), so that there is no stored energy. At $t = 0$ in this case, $i = i(0) = i_{tr}(0) + i_{ss}(0) = 0$, and therefore by equations (8), (10) and (11), $\Lambda = -E/R$. The complete solution is thus

$$i(t) = i_{tr} + i_{ss} = -\frac{E}{R}e^{st} + \frac{E}{R} = I(1-e^{st}) \tag{12}$$

where $I = E/R$ is the ultimate steady current.

(b) At $t = t_1$, $i = i(t_1) = I(1-e^{st_1})$. For $t > t_1$ $E = 0$, and therefore

$$L\,Di + Ri = 0 \qquad (t > t_1) \tag{13}$$

But while there is now no excitation, the circuit elements are not in a passive state: at the moment E vanishes there is the current $i(t_1)$ and the attendant energy, $\frac{1}{2}L[i(t_1)]^2$J, stored in L. The complete solution is now

$$i(t)_{(t > t_1)} = Ae^{s(t-t_1)} = i(t_1)e^{s(t-t_1)} = I(1-e^{st_1})e^{s(t-t_1)} \tag{14}$$

(c) For $0 < t < t_1$, noting that $s = -R/L$, equation (12) gives

$$v_L(t)_{(t < t_1)} = L\,Di = -Ls(E/R)e^{st} = Ee^{st} \tag{15}$$

and similarly for $t > t_1$, equation (14) gives

$$v_L(t)_{(t > t_1)} = L\,Di = Ls(E/R)(1-e^{st_1})e^{s(t-t_1)} = -E(1-e^{st_1})e^{s(t-t_1)} \tag{16}$$

As the bracketed term in equation (16) is always positive (since $s = -R/L$), there is an abrupt reversal in sign at the moment $t = t_1$.

Comment

(1) The natural mode, $s = -R/L$, has been described also as the *natural frequency*. In the sense that it has no vibratory connotations, it is an *imaginary frequency*. *Real frequency*, an attribute of a realistic vibration is, on the other hand, associated with an *imaginary mode*, $s = j\omega$. The general concept, embodying the notions both of decay and vibration or oscillation, is a *complex frequency* of form $s = \sigma + j\omega$, where, for a passive, linear network, σ is always negative. This is consistent with the decay in activity that must occur after removal of the excitation (the energy source), owing to the gradual dissipation of stored energy in the dissipative elements of the network. See Illustrations 2.8 to 2.12.

(2) Note that the excitation affects the transient component of the solution only in respect of its initial value: the form, Ae^{st}, is completely independent of the form of excitation; for it is the solution of the excitation-free or homogeneous differential equation. See Illustration 2.14.

ILLUSTRATION 2.2

The circuit in Figure 2.1 has been in a steady state for a long time with the switch S open. Derive a numerical expression for the current $i = i(t)$ in the inductor L at an instant t after the switch is closed.

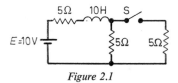

Figure 2.1

Plot to scale the current in the inductor and the voltage across it for a period from 1 s before to 2.5 s after closure of the switch.

Interpretation

Let $t = 0$ denote the instant of closing S. For $t < 0$ the resistance in series with L is $R_1 = 10\ \Omega$, and for $t > 0$ it is $R_2 = 7.5\ \Omega$. Then for $t > 0$ the homogeneous equation is

$$(7.5 + 10D)i = 0 \qquad (t > 0) \tag{1}$$

and the characteristic equation is

$$7.5 + 10s = 0 \tag{2}$$

whence the transient component is

$$i = i_{\text{tr}(t>0)} = Ae^{st} = Ae^{-3t/4} \tag{3}$$

The steady-state components of current before and after closing S at $t = 0$ are

$$i_{\text{ss}(t<0)} = E/R_1 = 10/10 = 1 \text{ A} \tag{4}$$
$$i_{\text{ss}(t>0)} = E/R_2 = 10/7.5 = 4/3 \text{ A} \tag{5}$$

The actual current at any instant t after closing S is

$$i = i(t)_{(t>0)} = (i_{\text{tr}} + i_{\text{ss}})_{(t>0)} = Ae^{-3t/4} + 4/3 \tag{6}$$

But the circuit was initially in the first steady state, impinging on the boundary $t = 0$, and therefore at $t = 0$, $i = i(0) = i_{\text{ss}(t<0)} = 1 \text{ A}$. Then, putting $t = 0$ in equation (6), $1 = A + 4/3$, $A = -1/3$, and

$$i = i(t)_{(t>0)} = \tfrac{4}{3}[1 - \tfrac{1}{4}e^{-3t/4}] \tag{7}$$

The voltage fall across L in the sense of i is

$$v_L = v_L(t)_{(t>0)} = L\frac{\mathrm{d}i}{\mathrm{d}t} = \frac{5}{2}e^{-3t/4} \tag{8}$$

Equations (7) and (8) are plotted in Figure 2.2.

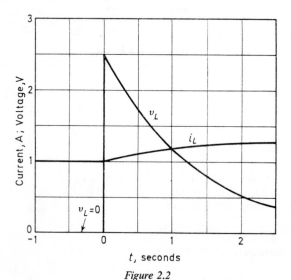

Figure 2.2

Comment

Figure 2.2 demonstrates the disturbance to an existing steady state due to an abrupt change in the value of a circuit element. Such a disturbance can, however, be regarded as one superimposed on the existing state, and can alternatively be attributed to the independent action of an abruptly applied excitation in the modified circuit, equal in magnitude but opposite in sign to the potential fall that would have existed in the original state across an element identical with the change made. This is an example of the *compensation theorem*.

The original steady current was $i_{ss(t<0)} = 1$ A, and the change in circuit resistance on closing S was $\Delta R = -2.5\ \Omega$. The equivalent independent excitation in the modified circuit is therefore a step of amplitude $\Delta E = -\Delta R i_{ss(t<0)} = -(-2.5) = 2.5$ V, acting in the same sense as $i_{ss(t<0)}$, and the response due to this excitation alone in the modified circuit may be quoted as

$$i'(t)_{(t>0)} = \frac{\Delta E}{R_2}[1-e^{-R_2 t/L}]$$

$$= \frac{2.5}{7.5}[1-e^{-3t/4}] = \frac{1}{3}[1-e^{-3t/4}]$$

The total current $i(t)_{(t>0)}$ is $i'(t)_{(t>0)}$ superimposed on the original steady current $i_{ss(t<0)}$, and therefore

$$i(t)_{(t>0)} = i_{ss(t<0)}+i'(t)$$
$$= 1+\tfrac{1}{3}[1-e^{-3t/4}]$$
$$= \tfrac{4}{3}[1-\tfrac{1}{4}e^{-3t/4}]$$

as before.

ILLUSTRATION 2.3

The switch S in Figure 2.1, Illustration 2.2, is reopened after remaining closed for (a), 4/3 s and (b), a sufficiently long time for the inductor current $i = i(t)$ to have reached its second steady-state value of 4/3 A. Obtain the corresponding numerical expressions for i at an instant t after S is opened.

Interpretation

Let $t = t_1$ denote the instant of opening S. For $t < t_1$ the circuit resistance is $R_2 = 7.5\ \Omega$, but at $t = t_1$ it reverts abruptly to the value $R_1 = 10\ \Omega$.

(a) The initial current at $t = t_1$ is the current at the instant $4/3$ s after S was closed in Illustration 2.2, or

$$i(t_1) = \tfrac{4}{3}[1 - \tfrac{1}{4}e^{-3t_1/4}] = \tfrac{4}{3} - \tfrac{1}{3}e^{-1} \tag{1}$$

By inspection, the steady-state current with S reopened is

$$i_{ss(t > t_1)} = E/R_1 = 10/10 = 1 \text{ A} \tag{2}$$

The homogeneous equation for $t > t_1$ is $(10 + 10D)i = 0$, the characteristic equation is $1 + s = 0$, and the transient component of the current is therefore

$$i = i_{tr(t > t_1)} = Ae^{st} = Ae^{-t} \tag{3}$$

The actual current at any instant $t > t_1$ is

$$i = i(t)_{(t > t_1)} = (i_{tr} + i_{ss})_{(t > t_1)} = (Ae^{-t} + 1)_{(t > t_1)} \tag{4}$$

Putting $t = 0$ in equation (4) and equating to the initial value given by equation (3) then gives $A = (1 - e^{-1})/3$, and therefore

$$i(t)_{(t > t_1)} = \tfrac{1}{3}(1 - e^{-1})e^{-t} + 1 = 1 + 0 \cdot 211e^{-t} \tag{5}$$

(b) Only the initial condition is different. In this case, $i(t_1) = i_{ss(t < t_1)} = 4/3$ A. Putting $t = 0$ in equation (4) and equating to this initial value then gives $A = 1/3$ and

$$i(t)_{(t > t_1)} = 1 + \tfrac{1}{3}e^{-t} \tag{6}$$

Comment

The compensation principle demonstrated in Illustration 2.2 is not directly applicable to case (a), since the current itself is still a function of time at the moment S is reopened. But it is applicable to case (b), as the initial current is entirely a steady-state value. The equivalent independent excitation in the modified circuit is a step of amplitude

$$\Delta E = -\Delta R i_{ss(t < t_1)} = -2 \cdot 5 \times \tfrac{4}{3} = -\tfrac{10}{3} \text{ V}$$

The response of the modified circuit to ΔE for $t > t_1$ is

$$i'(t)_{(t > t_1)} = \frac{\Delta E}{10}[1 - e^{-t}]$$

$$= -\tfrac{1}{3} + \tfrac{1}{3}e^{-t}$$

The total current for $t > t_1$ is then

$$i(t)_{(t > t_1)} = i_{ss(t < t_1)} + i'(t)_{(t > t_1)}$$

$$= 1 + \tfrac{1}{3}e^{-t}$$

as before.

ILLUSTRATION 2.4

Two capacitors C_1 and C_2 have a common negative terminal. Capacitor C_1 of 5 μF is charged to $+100$ V and capacitor C_2 of 20 μF is charged to $+50$ V. The positive terminals are suddenly connected together through a resistor R of 2 MΩ. Find

(1) the final voltage reached,
(2) the total energy dissipated in R,
(3) the total charge transferred through R.

Derive an expression for the voltage across C_1 as a function of time, and find the time taken for this voltage to decay to 90 V.

(L.U. Part 2, Electrical Theory and Measurements).

Interpretation

Let $q_1(0)$, $q_2(0)$ denote the initial charges at $t = 0$, when the circuit is completed through R. Then, $v_1(0) = 100 = q_1(0)/C_1$, $v_2(0) = 50 = q_2(0)/C_2$, and for a current i in the sense of $v_1(0)$, $\sum v = 0$ when

$$Ri + \frac{q_1}{C_1} + \frac{q_2}{C_2} - \frac{q_1(0)}{C_1} + \frac{q_2(0)}{C_2} = 0$$

or

$$Ri + \frac{q_1}{C_1} + \frac{q_2}{C_2} = 50$$

Since i is common to all elements in the circuit, $dq_1/dt = dq_2/dt$, and differentiation gives

$$R \frac{di}{dt} + \left(\frac{1}{C_1} + \frac{1}{C_2} \right) i = 0$$

or, substituting values,

$$2 \frac{di}{dt} + \frac{1}{4} i = 0$$

The complete solution is

$$i = Ae^{st} = Ae^{-t/8} = \frac{50}{2 \times 10^6} \cdot e^{-t/8} = 25e^{-t/8} \text{ μA}$$

(1) The capacitor voltages become equal as $t \to \infty$ and $i \to 0$. Let V denote the final common voltage. Then the total final stored energy is

$$w = \tfrac{1}{2} V^2 (C_1 + C_2) = 12 \cdot 5 \times 10^{-6} V^2 \text{ J}$$

The initial stored energy is

$$w(0) = \tfrac{1}{2}[C_1 v_1^2(0) + C_2 v_2^2(0)] = 0.05 \text{ J}$$

The energy dissipated as C_1 discharges into C_2 is

$$w_d = \int_0^\infty i^2 R \, dt = 0.005 \text{ J}$$

and the total final stored energy is therefore

$$w = w(0) - w_d = 0.045 = 12.5 \times 10^{-6} V^2 \text{ J}$$

from which

$$V = 10 \sqrt{\left(\frac{450}{12.5}\right)} = 60 \text{ V}$$

(2) The total energy dissipated in R is 0.005 J as found above
(3) The total charge transferred through R is

$$q_T = \int_0^\infty i \, dt = 200 \ \mu\text{C}$$

The voltage across C_1 is

$$v_1 = \frac{q_1(0)}{C_1} - \frac{q_1}{C_1} = \frac{q_2(0)}{C_2} + \frac{q_2}{C_2} + Ri$$

$$= 50 + \frac{1}{C_2} \int_0^t i \, dt + Ri$$

$$= 50 + \frac{25}{20} \left[-8e^{-t/8}\right]_0^t + 50e^{-t/8}$$

$$= 60 + 40e^{-t/8}$$

When $v_1 = 90$ V,

$$40e^{-t/8} = 30$$

whence

$$t = 2.32 \text{ s}$$

Comment

(1) The final voltage derived from energy considerations is consistent with the KVL equation and with the transfer of charge.
(a) The solution $v_1 = 60 + 40e^{-t/8}$ approaches 60 V as $t \to \infty$; but as $t \to \infty$, $i \to 0$, so that this is also the voltage across C_2.

(b) $\displaystyle\int_0^\infty i\,\mathrm{d}t = 200\ \mu\mathrm{C}$ checks with the charge lost by C_1 in falling in potential through 40 V, for $C_1\Delta v = 5\times10^{-6}\times40 = 200\ \mu\mathrm{C}$.

(c) The initial charge in C_2 is $C_2v(0) = 10^{-3}$ C. The charge transferred from C_1 is $0{\cdot}2\times10^{-3}$ C, so that the ultimate total charge in C_2 is $1{\cdot}2\times10^{-3}$ C, and $V = 1{\cdot}2\times10^{-3}/C_2 = 60$ V.

(2) While there is a change in stored energy, the charge leaving the one capacitor equals that entering the other: the moving charge, like the current, is common to the system; and it could have been assumed at the onset that, although $q_1(0) \neq q_2(0)$, the instantaneous charges q_1 and q_2 are equal.

ILLUSTRATION 2.5

Figure 2.3 shows a 100 pF capacitor C arranged to discharge through a 10 kΩ resistor R and a Zener diode Z. The Zener diode maintains a constant voltage drop of 6 V across it until the current falls to 5 μA; for smaller currents than this the diode may be regarded as an open circuit.

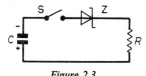

Figure 2.3

The capacitor is charged to 11 V before the switch S is closed. Determine the time taken for the diode to halt the discharge. What will then be the voltages across the capacitor?

(L. U. Part 2, Electrical Theory and Measurements)

Interpretation

The Zener diode, while maintaining the constant voltage, is indistinguishable from an emf of 6 V having a polarity opposing that of the capacitor voltage. This equivalent emf is abruptly replaced by an open-circuit when

$$i = \frac{11-6}{R}\ e^{-t/CR} = \frac{5}{10^4}\ e^{-10^6 t} = 5\times10^{-6}\mathrm{A}$$

whence $t = 4{\cdot}6\ \mu$s.

When at $t = 4{\cdot}6\ \mu$s the current has fallen to the critical value of 5 μA, the capacitor stops discharging since the diode (as here idealised) then abruptly opens the circuit. Let v_c denote the capacitor voltage at this moment. Then

$$\frac{v_c - 6}{R} = \frac{v_c - 6}{10^4} = 5 \times 10^{-6}$$

or

$$v_c - 6 = 5 \times 10^{-2}$$

whence

$$v_c = 6 \cdot 05 \text{ V}$$

Comment

The Zener diode is a type of semi-conductor diode characterised by an almost constant voltage-drop over a range of its reverse voltage-current characteristic. It is widely used as a voltage stabilising device.

This problem is important as an illustration of the kind of point-by-point calculations that can be made in electronic switching circuits by the idealisation of electronic elements. Such idealisation is usually justified in practice as a fair approximation.

ILLUSTRATION 2.6

A current-source $i(t) = U(t)I$ having a shunt conductance G is connected in parallel with (a), a pure capacitor C; and (b), a pure inductor. L. Show that these circuits are, respectively, the duals of an emf source $e(t) = U(t)E$ having a series resistance R applied to (c), a pure inductor L; and (d), a pure capacitor C. Hence quote appropriate solutions by applying the principle of duality to forms of solution already encountered.

When the current-source $U(t) I$ in association with the shunt conductance G is transformed into an equivalent emf source with series resistance, circuits (a) and (b) assume the forms of circuits (d) and (c), respectively. Distinguish between this transformation and the concept of duality.

Interpretation

(1) Consider the familiar relations,

$$v_c = \frac{q}{C} = \frac{1}{C} \int_0^t i_c \, dt$$

and

$$v_L = \frac{d}{dt} \Lambda = \frac{d}{dt} Li_L = L \frac{di_L}{dt}$$

(where Λ denotes flux-linkage).
By differentiating the first,

$$i_c = \frac{dq}{dt} = C \frac{dv_c}{dt}$$

By integrating the second,

$$i_L = \frac{1}{L} \int_0^t v_L \, dt$$

The parallelisms

$$v_c = \frac{1}{C} \int_0^t i_c \, dt \quad \text{and} \quad i_L = \frac{1}{L} \int_0^t v_L \, dt$$

$$v_L = L \frac{di}{dt} \quad \text{and} \quad i_c = C \frac{dv_c}{dt}$$

are such that a systematic interchange of symbols transforms an expression for voltage in the one case into current for the other, and vice-versa. Capacitance and inductance are therefore dual parameters.

(2) The given circuits are shown in Figures 2.4(a) and (b).
For (a),

$$Gv + C \frac{dv}{dt} = I \quad (t > 0)$$

For (b),

$$Gv + \frac{1}{L} \int_0^t v \, dt = I \quad (t > 0)$$

The duals of these equations, given by substituting the dual parameters, are respectively

$$Ri + L \frac{di}{dt} = E$$

and

$$Ri + \frac{1}{C} \int_0^t i \, dt = E$$

The dual circuits conforming to these equations and corresponding to those in Illustrations 2.1 and 2.4 are shown in Figures 2.4(c) and (d).

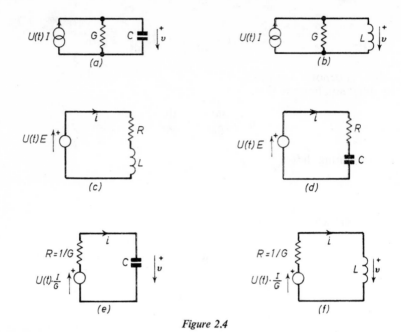

Figure 2.4

The complete solutions for these are known, and by the principle of duality, the solutions for Figures 2.4(a) and (b) are given merely by exchanging symbols. They are therefore

(a) $v = v_{\text{tr}} + v_{\text{ss}} = \dfrac{I}{G}(1 - e^{-Gt/C})$

(b) $v = v_{\text{tr}} + 0 = \dfrac{I}{G} \cdot e^{-t/LG}$

(3) The transformed circuits are shown in Figures 2.4(e) and (f). Circuit (e) is now the series equivalent of (a), but no longer the dual of (d). This means that the solutions for (a) and (b) may be obtained alternatively by solving (e) and (f) for i as a function of $U(t)E$, where $E = I/G$, but not by a mere exchange of symbols in these solutions. In (e), for example, the solution for $t > 0$ is

$$i = \frac{E}{R}e^{-t/CR} = \frac{I}{G} \cdot Ge^{-Gt/C} = Ie^{-Gt/C}$$

and

$$v = E - iR = E(1 - e^{-t/CR})$$

$$= \frac{I}{G}(1 - e^{-Gt/C})$$

This is the correct solution for v in (a); but it does not correspond to merely exchanging symbols in the solution $i = Ie^{-Gt/C}$, which is the capacitor current $(C\,dv/dt)$ in (a).

Comment

An equivalent circuit is one having a configuration and set of elements different from those of a given circuit, but giving at a particular point the same response to the same excitation at another particular point. But a dual circuit is one for which the pattern of the nodal-voltage (or loop-current) equations is exactly the same as the pattern of the loop-current (or nodal-voltage) equations for a given circuit, so that a systematic interchange of coefficients and variables transforms the one set into the other. The dual network requires a dual configuration (topological graph) as well as dual elements.

ILLUSTRATION 2.7

A current step of form $U(t)I$ is applied to a resistor R in parallel with a capacitor C. Find the rise-time T_r for the voltage across the circuit from 10% to 90% of its final steady value, and show that it is related to the half-power or 3 dB frequency ω_2 for steady sinusoidal current excitation, by $\omega_2 = 2 \cdot 2/T_r$.

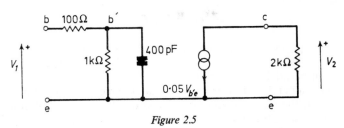

Figure 2.5

Figure 2.5 is a circuit model for a transistor amplifier at high frequencies. Determine directly, or in terms of the rise-time T_r, the upper 3dB frequency in the response V_2 to a sinusoidal excitation V_1, and V_2/V_1 at this frequency.

(P.C.L., B.Sc. (Hons), CNAA, Electronics, Year 2)

Interpretation

(1) The nodal-voltage equation for the stated arrangement is

$$Gv + CDv = I \qquad (t > 0) \tag{1}$$

where $G = 1/R$.

6

Equation (1) may be solved formally by putting $I = 0$ to yield a homogeneous equation whose solution has the form $v = Ae^{st}$, where $s = -G/C$. The steady-state solution is quotable as $v_{ss} = IR = I/G$, and for C initially passive, $v(0) = v_{tr}(0) + v_{ss}(0) = 0$. This gives $A = -I/G$, and therefore

$$v(t) = v_{tr} + v_{ss} = \frac{I}{G}(1 - e^{-Gt/C}) \tag{2}$$

Alternatively, equation (2) might be quoted as the dual of the solution for the case of an excitation $U(t)E$ applied to an inductor in series with a resistor (see Illustration 2.1).

The rise-time from 10% to 90% of the final steady voltage is given by solving the equations

$$0\cdot1 = 1 - e^{-Gt_1/C} \quad \text{and} \quad 0\cdot9 = 1 - e^{-Gt_2/C} \tag{3}$$

for $t_2 - t_1$. The solution is

$$T_r = t_2 - t_1 = 2\cdot2 C/G \tag{4}$$

In Illustration 4.5 the half-power frequency is shown to occur when $X = R$ for a series circuit, or $B = G$ for a parallel circuit. In this case $B = \omega_2 C$, and when $B = G$, $C/G = 1/\omega_2$ so that equation (4) may be put in the form

$$\omega_2 = 2.2/T_r \tag{5}$$

(2) That part of the input circuit comprising V_1 and the two resistors may be transformed into a current source $I_1 = V_1/100$ acting from e to b', in parallel with $G_T = 10^{-2} + 10^{-3} = 1\cdot1 \times 10^{-2}$ S.

(a) The upper 3 dB frequency ω_2 is that at which the output power is halved and the output voltage is reduced to $1/\sqrt{2}$ of its low-frequency value. But the output voltage is proportional to $V_{b'e}$, and ω_2 therefore occurs when $G_T = \omega_2 C$, where $C = 400$ pF. Substituting values, $\omega_2 = 1\cdot1 \times 10^8/4 = 2\cdot75 \times 10^7$, or $f_2 = 4\cdot38$ MHz.

(b) At a frequency low enough for C to be ignored,

$$V_{b'e} = V_1 \times 10^3/1\cdot1 \times 10^3 = V_1/1\cdot1$$

Then,

$$V_2 = -0\cdot05 V_{b'e} \times 2 \times 10^3 = -100 V_1/1\cdot1$$

and

$$V_2/V_{1(LF)} = -100/1\cdot1 = -91$$

Therefore at ω_2,

$$V_2/V_{1(\omega_2)} = -91/\sqrt{2}$$

Comment

(1) The correlation shown to exist between responses in the time and frequency domains is of great practical importance: the bandwidth of a network or amplifier can be estimated quickly from an oscilloscope trace of its response to a periodic rectangular waveform whose period is long enough to simulate a step (square-wave testing).

There is, however, a general correlation between responses in the time and frequency domains, expressed through the Fourier transforms

$$f(t) = \frac{1}{2\pi} \int_{-\infty}^{+\infty} F(\omega)e^{j\omega t} \, d\omega \qquad (6)$$

and

$$F(\omega) = \int_{-\infty}^{+\infty} f(t)e^{-j\omega t} \, dt \qquad (7)$$

which can be derived from the Fourier series.

A unit step function $U(t)$, as involved in this illustration, is expressed by the Fourier integral

$$U(t) = \frac{1}{2} + \frac{1}{\pi} \int_{0}^{\infty} \frac{\sin \omega t}{\omega} \, d\omega \qquad (8)$$

Now consider the parallel CG circuit. Its voltage response $v(t)$ to a sinusoidal current excitation $I \sin \omega t$ may be put in the form

$$v(t) = I \left[\frac{G}{G^2 + \omega^2 C^2} \sin \omega t - \frac{\omega C}{G^2 + \omega^2 C^2} \cos \omega t \right] \qquad (9)$$

Then by equation (8) the response to a current step $U(t)I$ is

$$v(t) = \frac{I}{\pi} \left[\int_{0}^{\infty} \frac{G}{G^2 + \omega^2 C^2} \cdot \frac{\sin \omega t}{\omega} \, d\omega - \int_{0}^{\infty} \frac{\omega C}{G^2 + \omega^2 C^2} \cdot \frac{\cos \omega t}{\omega} \, d\omega \right]$$

$$+ \frac{I}{2G} = \frac{I}{G} \left[1 - e^{-Gt/C} \right] \qquad (10)$$

This result, derived from steady-state response in the frequency domain, is identical with equation (2), derived from the differential equation in the time domain.

(2) Figure 2.5 is obtained from the hybrid-π equivalent circuit by applying the relation $Y_{in} = Y + Y_f(1 - A)$ to $C_{b'e}$ and $C_{b'c}$. See Illustration 1.9. A is approximated by $-g_m R_2$, where R_2 is the load resistor (in this case 2 kΩ).

ILLUSTRATION 2.8

An excitation $e(t)$ is applied to a circuit comprising an inductor L of resistance R in series with a capacitor C. Find expressions for the natural frequencies in the response $i(t)$, and show that when $e(t) = Ee^{st}$, $i(t)$ is of the form Ie^{st} and that the driving-point impedance can be expressed as an explicit function of s and the natural frequencies. Confirm that when $s = j\omega$, as for sinusoidal excitation, this impedance function is equivalent to the usual a.c. expression, $Z = Z(j\omega) = R + j(\omega L - 1/\omega C)$.

Interpretation

The Kirchhoff voltage-law equation is

$$L\frac{di}{dt} + Ri + \frac{1}{C}\int_0^t i\, dt = e(t) \tag{1}$$

This integro-differential equation becomes a second-order differential equation when re-written in terms of charge q, since $i = dq/dt$. Then, in operational notation,

$$(LD^2 + RD + 1/C)q = e(t) \tag{2}$$

Setting $e(t) = 0$ then gives the excitation-free or homogeneous equation

$$[D^2 + (R/L)D + 1/LC]q = 0 \tag{3}$$

The solution of an nth. order homogeneous equation comprises n terms of the form $a_k e^{s_k t}$, and the solution in this case is therefore of the form

$$q = a_1 e^{s_1 t} + a_2 e^{s_2 t} \tag{4}$$

Substituting this solution into equation (3) and performing the differentiations then identifies the natural frequencies s_1, s_2 with the solutions of the characteristic equation

$$s^2 + (R/L)s + 1/LC = 0 \tag{5}$$

or,

$$(s - s_1)(s - s_2) = 0 \tag{6}$$

where

$$s_1, s_2 = -\frac{R}{2L} \pm \sqrt{\left\{\frac{R^2}{4L^2} - \frac{1}{LC}\right\}} \tag{7}$$

s_1 and s_2 may be real and negative (*imaginary frequencies*), or *complex conjugates* when $(R^2/4L^2) < 1/LC$. They are then *complex frequencies*, of the forms

$$s_1 = \sigma+j\omega_0, \quad s_2 = \sigma-j\omega_0 \tag{8}$$

where

$$\sigma = -R/2L, \quad \omega_0 = \sqrt{\{(1/LC)-R^2/4L^2\}} \tag{9}$$

The current is given as

$$i = \frac{dq}{dt} = s_1a_1e^{s_1t}+s_2a_2e^{s_2t} = A_1e^{s_1t}+A_2e^{s_2t} \tag{10}$$

When $e(t) = Ee^{st}$, equation (2) becomes, with slight rearrangement,

$$L[D^2+(R/L)D+1/LC]q = Ee^{st} \tag{11}$$

and the complete solution is given directly by substituting s for D, in the form

$$q = \frac{1}{F(s)} \cdot Ee^{st} = \frac{Ee^{st}}{(s^2+(R/L)s+1/LC)L} = Qe^{st} \tag{12}$$

and

$$i = \frac{dq}{dt} = \frac{sEe^{st}}{[s^2+(R/L)s+1/LC]L} = Ie^{st} \tag{13}$$

But the bracketed term in the denominator of equation (13) is identical in the form with the left hand side of equation (5), and with its factors in equation (6). Thus,

$$i = Ie^{st} = \frac{1}{L} \cdot \frac{s}{(s-s_1)(s-s_2)} \cdot Ee^{st} \tag{14}$$

and the driving-point impedance is

$$Z(s) = \frac{Ee^{st}}{Ie^{st}} = L \cdot \frac{(s-s_1)(s-s_2)}{s} \tag{15}$$

Equation (15) may be rewritten as

$$Z(s) = L[s-(s_2+s_1)+s_1s_2/s]$$
$$= L[s-2\sigma+(\sigma^2+\omega_0^2)/s]$$

where $2\sigma = -R/L$ and $\sigma^2+\omega_0^2 = 1/LC$, from equation (9). Then, for $s = j\omega$,

$$Z(j\omega) = L[j\omega+(R/L)+1/j\omega LC)]$$
$$= R+j\omega L-j/\omega C$$

which is the normal steady-state impedance to a sinusoidal excitation.

76

Comment

Equation (15) exemplifies an alternative way of stating the driving-point impedance of a network, in terms of its fixed natural frequencies and a single variable frequency, s. While for $s = j\omega$ the expression is equivalent, as shown, to the usual complex ac impedance formula, the exclusion of circuit elements (except for the constant multiplier $1/L$) makes the form ideal for the study of behaviour as a function of frequency. It is, however, not restricted to the sinusoidal steady state, but serves also for finding transient response to a time-varying excitation that has been temporarily transformed into a complex-frequency function by Laplace transformation: s is then a complex-frequency variable of form $s = \sigma + j\omega$. See Illustration 2.12.

ILLUSTRATION 2.9

(1) Obtain for Figure 2.6 expressions for the natural frequencies in the potential v_1 of node A relative to B in response to $e_1(t)$. Show that these frequencies can never be complex, and confirm that i_2 has the same frequencies.

(2) Obtain numerical expressions for $v_1 = v_1(t)$ and $i_2 = i_2(t)$ when $R_1 = R_2 = R = 1\,\Omega$, $C_1 = C_2 = C = 1\,\text{F}$ and $e_1(t)$ is a voltage step $U(t)E_1$, where $E_1 = 1\,\text{V}$. C_1 and C_2 are initially uncharged.

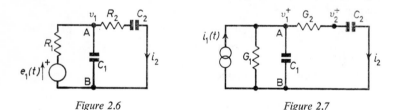

Figure 2.6   Figure 2.7

Interpretation

(1) Nodal voltage analysis is advantageous. In Figure 2.7, $i_1(t) = e_1(t)/R_1$ in parallel with $G_1 = 1/R_1$ replaces $e_1(t)$ in series with R_1, and the junction of C_2 with $G_2 = 1/R_2$ is treated as a node of potential v_2.

Denoting d/dt by the operator D,

$$(G_1 + G_2 + C_1 D)v_1 - G_2 v_2 = i_1(t) \tag{1}$$

$$(G_2 + C_2 D)v_2 - G_2 v_1 = 0 \tag{2}$$

Substituting from (2) into (1), manipulating and putting $i_1(t) = 0$, gives the homogeneous equation

$$(D^2 + a_1 D + a_0)v_1 = 0 \qquad (3)$$

and the characteristic equation

$$s^2 + a_1 s + a_0 = 0 \qquad (4)$$

where

$$a_1 = \frac{G_1}{C_1} + \frac{G_2}{C_2} + \frac{G_2}{C_1} \quad \text{and} \quad a_0 = \frac{G_1 G_2}{C_1 C_2} \qquad (5)$$

The natural frequencies are

$$s_1, s_2 = -\tfrac{1}{2}a_1 \pm \tfrac{1}{2}\sqrt{(a_1^2 - 4a_0)} \qquad (6)$$

They can never be complex if $a_1^2 > 4a_0$ for all finite, positive circuit values. Let $G_1/C_1 = p$, $G_2/C_2 = q$, and $G_2/C_1 = r$. Then,

$$\begin{aligned}
a_1^2 - 4a_0 &= (p+q+r)^2 - 4pq \\
&= p^2 - 2pq + q^2 + (r^2 + 2pr + 2qr) \\
&= (p-q)^2 + (r^2 + 2pr + 2qr)
\end{aligned}$$

But for real, positive circuit elements, $(p-q)^2$ and all other terms are always real and positive. Hence $a_1^2 - 4a_0$ is always positive and the root is always real.

Eliminating instead v_1 from equations (1) and (2) gives the homogeneous equation

$$(D^2 + a_1 D + a_0)v_2 = 0$$

where a_1 and a_0 are still defined by equation (5). Hence the same characteristic equation, equation (4), defines the natural frequencies for both v_1 and v_2, and also for i_2, since $i_2 = G_2(v_1 - v_2)$.

(2) When $R_1 = R_2 = R$ and $C_1 = C_2 = C$, equation (6) becomes $s_1, s_2 = G(-3 \pm \sqrt{5})/2C$, and for $R = 1\,\Omega$ and $C = 1$ F,

$$s_1 = -0\cdot382, \quad s_2 = -2\cdot618 \qquad (7)$$

In Figure 2.7,

$$i_1(t) = U(t)E_1/R_1 = U(t)I_1$$

where $I_1 = 1A$. By inspection, the steady-state components of the solutions are

$$v_{1ss} = I_1/G_1 = 1, \quad i_{2ss} = 0, \quad v_{2ss} = 1 \qquad (8)$$

and the complete solutions for v_1 and v_2 therefore have the forms

$$v_1 = v_1(t) = v_{1ss} + v_{1tr} = 1 + A_1 e^{s_1 t} + A_2 e^{s_2 t} \qquad (9)$$

$$v_2 = v_2(t) = v_{2ss} + v_{2tr} = 1 + B_1 e^{s_1 t} + B_2 e^{s_2 t} \qquad (10)$$

where $A_1\,A_2\,B_1\,B_2$ are governed by the initial conditions.

As C_1 and C_2 are uncharged initially, $q_1(0) = 0$, $q_2(0) = 0$, $v_1(0) = 0$ and $v_2(0) = 0$. Putting $t = 0$ in equation (9) then gives

$$A_1 + A_2 = -1 \tag{11}$$

A further initial condition is provided by the charging current, which equals the source current I_1 at $t = 0$ (an uncharged capacitor behaves momentarily like a short-circuit). By differentiating equation (9), setting $t = 0$, and equating $C_1 \, dv_1/dt$ to I_1, is obtained

$$-0.382A_1 - 2.618A_2 = 1 \tag{12}$$

From equations (11) and (12), $A_1 = -0.724$, $A_2 = -0.276$ and

$$v_1 = v_1(t) = 1 - 0.724e^{-0.382t} - 0.276e^{-2.618t} \tag{13}$$

The initial condition $v_2(0) = 0$ gives similarly from equation (10),

$$B_1 + B_2 = -1 \tag{14}$$

But as the initial value of $i_2 = C_2 \, dv_2/dt$ is zero (since both v_1 and v_2 are initially zero), differentiation of equation (10) leads, for $t = 0$, to

$$-0.382B_1 - 2.618B_2 = 0 \tag{15}$$

From equations (14) and (15), $B_1 = -1.171$, $B_2 = 0.171$ and

$$v_2 = v_2(t) = 1 - 1.171e^{-0.382t} + 0.171e^{-2.618t} \tag{16}$$

Then,

$$i_2 = i_2(t) = G_2[v_1(t) - v_2(t)] = 0.447(e^{-0.382t} - e^{-2.618t}) \tag{17}$$

Comment

(a) Nodal-voltage analysis has advantageously excluded integrals from the Kirchhoff-law equations, since for a capacitor $i = C dv/dt$.

(b) The natural frequencies are common to all responses throughout the circuit: it is only the constants associated with particular responses that differ, as exemplified by A_1, A_2 and B_1, B_2. As many initial conditions are required to find the constants as there are natural frequencies.

(c) The natural frequencies for a linear, passive network with one kind of storage element only (C or L) are always imaginary, corresponding to exponents that are real and negative.

ILLUSTRATION 2.10

The amplifier included in Figure 2.8 is direct coupled and has a real, positive, open-circuit voltage gain A. Negligible current traverses its input terminals under transitory or steady conditions of excitation, and negligible loading is imposed on its output terminals by the external circuit (in steady-state terms the input impedance may be regarded as infinite and the output impedance as negligible).

(1) Obtain expressions for the natural frequencies in $v_1(t)$ and $v_2(t)$ in response to the excitation $i_1(t)$ when $R_1 = R_2 = R$ and $C_1 = C_2 = C$, and determine the frequencies for $A = 0, 1, 2$ and 3 when R and C have the scaled values $R = 1\Omega, C = 1$ F.

(2) Obtain numerical expressions for $v_1(t)$ when $i_1(t)$ is a unit current-step $U(t)$, and the voltage-gain is $A = 2$ and $A = 3$.

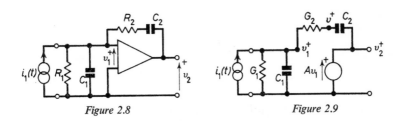

Figure 2.8   Figure 2.9

Interpretation

An equivalent circuit styled for a nodal-voltage analysis is shown in Figure 2.9. The amplifier is represented with its input terminal-pair internally open-circuited, which is consistent with zero input current for any excitation $v_1 = v_1(t)$; and with its output terminal-pair fed from a constant emf Av_1 equal to the open-circuit voltage, which is consistent with negligible loading by the external circuit. $G_1 = 1/R_1$ and the junction of $G_2 = 1/R_2$ with C_2 is treated as a node of potential v.

(1) Denoting d/dt by the operator D,

$$(G_1+G_2+C_1D)v_1-G_2v = i_1(t) \tag{1}$$

$$(G_2+C_2D)v-G_2v_1-C_2Dv_2 = 0 \tag{2}$$

Putting $G_1 = G_2 = G$, $C_1 = C_2 = C$, $v_2 = Av_1$, eliminating v and putting $i_1(t) = 0$ gives the excitation-free or homogeneous equation

$$\{D^2+[(3-A)G/C]D+G^2/C^2\}v_1 = 0 \tag{3}$$

and the characteristic equation

$$s^2 +[(3-A)G/C]s+G^2/C^2 = 0 \tag{4}$$

having roots

$$s_1, s_2 = [(G/2C][(A-3)\pm\sqrt{\{(3-A)^2-4\}]} \tag{5}$$

which are the natural frequencies for the circuit.

When
$$G = 1/R = 1 \text{ S} \quad \text{and} \quad C = 1 \text{ F},$$
$$s_1, s_2 = -\tfrac{1}{2}(3-A) \pm \tfrac{1}{2}\sqrt{[(3-A)^2-4]} \qquad (6)$$

These natural frequencies are tabulated below as functions of A.

A	s_1	s_2
0	$-0 \cdot 382$	$-2 \cdot 618$
1	$-1 \cdot 0$	$-1 \cdot 0$
2	$-\tfrac{1}{2}+j\sqrt{3}/2$	$-\tfrac{1}{2}-j\sqrt{3}/2$
3	$j\,1 \cdot 0$	$-j\,1 \cdot 0$

(2) The complete solution for v_1 has the form

$$v_1 = v_1(t) = v_{1ss} + K_1 e^{s_1 t} + K_2 e^{s_2 t} \qquad (7)$$

The steady-state and the initial conditions are similar to those in Illustration 2.9. Since $i_1(t) = U(t)$, $i_1 = 0$, $t < 0$ and $i_1 = 1$, $t > 0$. Then by inspection,

$$v_{1ss} = I_1/G_1 = 1 \text{ V} \qquad (8)$$

and from the initial conditions,

$$K_1 + K_2 = -1 \quad \text{and} \quad s_1 K_1 + s_2 K_2 = 1 \qquad (9)$$

which yield

$$K_1 = (1+s_2)/(s_1-s_2) \quad \text{and} \quad K_2 = -(1+K_1) \qquad (10)$$

Solutions to equation (7), for the natural frequencies when $A = 2$ and $A = 3$, are:

$$A = 2, \quad v_1(t) = 1 + \frac{2}{\sqrt{3}} e^{-t/2} \sin\left(\frac{\sqrt{3}}{2} t - \frac{\pi}{3}\right) \qquad (11)$$

$$A = 3, \quad v_1(t) = 1 + \sqrt{2}\sin(t - \pi/8) \qquad (12)$$

Comment

The circuit is the same as Figure 2.6, Illustration 2.9, except for the inclusion of an amplifier. This modification gives control over the coefficient of D in equation (3), so that the natural frequencies are not constrained to be imaginary by roots of a characteristic equation that can only be real and negative, but may be complex or even real, as when $A = 3$ and the roots of equation (4) are wholly imaginary (*real frequency* corresponds to $s = j\omega$).

When the natural frequencies are complex, the transient component of the complete solution is a damped sinusoidal oscillation, as exemplified by equation (11) for the case $A = 2$, in which the decay is governed by the factor $e^{-t/2}$.

The case $A = 3$ is of special interest, for the coefficient of D in equation (3) is then zero, the roots of equation (4) are wholly imaginary, and the transient component in the complete solution, equation (12), has no decay factor. This implies that a continuous state of sinusoidal oscillation may exist in absence of an external excitation: $A = 3$ is thus the critical amplification at the threshold of continuous self-oscillation, at which an energy-balance is preserved.

The threshold criterion for continuous oscillation is thus either that for which the natural frequencies just become real and just fall on the $j\omega$-axis in the complex-frequency plane (see Illustration 2.11), or that for which the decay factor α is just zero, in a solution of the form $\int(t) = Ae^{\alpha t} \sin(\omega t + \phi)$, since $\alpha = 0$ is consistent with $\sigma = 0$ in $s_1, s_2 = \sigma \pm j\omega_0$. These transient or time-domain criteria have equivalents in steady-state frequency-domain approaches (see Illustrations 6.7, 6.8, 6.9 and 6.10).

ILLUSTRATION 2.11

Define the terms zero and pole as applied to an immitance (impedance or admittance) function.

A circuit, scaled to impedance and frequency references of $1\,\Omega$ and 1 rad/s (see Chapter 3) comprises an inductor of resistance R and inductance 1 H, in series with a 1 F capacitor. Plot to scale in the complex-frequency plane the pole-zero locations for the admittance for $R = 2$, 1, and zero ohms. For $R = 1\,\Omega$, use the pole-zero diagram to find graphically the variation in magnitude and phase of the current in response to a sinusoidal emf of 1 V as its frequency is varied from $\omega = 0$ to $\omega = 2 \cdot 0$ rad/s.

Interpretation

(a) In Illustration 2.8, the immitance of an LCR series circuit to an excitation Ee^{st} is interpreted as an impedance

$$Z(s) = L \cdot \frac{(s-s_1)(s-s_2)}{s}$$

or as an admittance

$$Y(s) = \frac{1}{L} \cdot \frac{s}{(s-s_1)(s-s_2)}$$

where s_1, s_2 are the natural frequencies.

When $s = s_1$ or $s = s_2$, $Z(s)$ becomes zero and $Y(s)$ infinite; and when $s = 0$, $Z(s)$ becomes infinite and $Y(s)$ zero. Each value of s in the factored numerator for which the function becomes zero is called a *zero*; and each value in the factored denominator for which the denominator becomes zero and the function infinite is called a *pole*. The natural frequencies satisfying the homogeneous equation $F(D)i = 0$ are therefore the poles of $Y(s)$ and the zeros of $Z(s)$. In this case, $s = 0$ is a zero of $Y(s)$ and a pole of $Z(s)$. Only when the circuit resistance R is zero, so that $\sigma_1, \sigma_2 = 0$ and $s_1, s_2 = \pm j\omega_0$, can s_1 or s_2 be cancelled by a term of form $\pm j\omega_0$ derived from a sinusoidal excitation of phasor form $Ee^{j\omega t}$.

(b) The natural frequencies

$$s_1, s_2 = -\frac{R}{2L} \pm \sqrt{\left\{ \frac{R^2}{4L^2} - \frac{1}{LC} \right\}}$$

are complex for $R < 2\,\Omega$, and then have the forms

$$s_1, s_2 = \sigma \pm j\omega_0 = -\frac{R}{2L} \pm j \sqrt{\left\{ \frac{1}{LC} - \frac{R^2}{4L^2} \right\}}$$

They are tabulated below and located in the complex-frequency plane in Figure 2.10(a) as the poles of $Y(s) = s/L(s-s_1)(s-s_2)$. A zero (circle) is also included at $s = 0$.

R	σ	ω_0	s_1	s_2
2	-1	0	-1	-1
1	$-\frac{1}{2}$	$\sqrt{3}/2$	$-\frac{1}{2}+j\sqrt{3}/2$	$-\frac{1}{2}-j\sqrt{3}/2$
0	0	1	$j\,1$	$-j\,1$

Figure 2.10(b) is the pole-zero diagram for $R = 1\,\Omega$. The lines from the typical point $s = j0.5$ to the zero at the origin and to the poles s_1, s_2 are phasors representing s, $s-s_1$ and $s-s_2$ for a real frequency $\omega = 0.5$ rad/s. Since $E = 1$ V and $L = 1$ H, $I(s) = EY(s) = s/(s-s_1)(s-s_2)$; and the phasor lengths to scale therefore give directly the magnitude of the current, while their inclinations to the real axis give its phase. Measurements on Figure 2.10(b) drawn to a

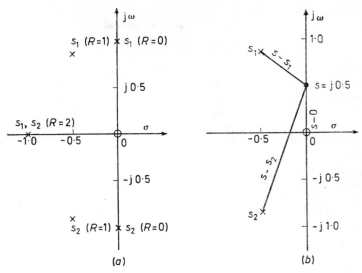

Figure 2.10

scale of 1 cm = 0·2 rad/s give

$$|I(j\omega)|_{\omega = 0.5} = \frac{|j\omega|}{|j\omega - s_1| \cdot |j\omega - s_2|} = \frac{2\cdot5 \text{ cm}}{3\cdot1 \text{ cm} \times 7\cdot25 \text{ cm}} \times 5 = 0.556$$

and

$$\theta = \arg I(j\omega)_{\omega = 0.5} = \arg j\omega - [\arg (j\omega - s_1] + \arg (j\omega - s_2)]$$

$$= 90° - (-36° + 70°) = 56°$$

Currents for other frequencies are estimated similarly, and are tabulated below for the range $\omega = 0$ to $\omega = 2$ rad/s. From these a resonance curve may be sketched.

ω, rad/s	I, A	$\theta = \arg I$, deg
0	0	$\to 90, \omega \to 0$
0·25	0·3	75
0·5	0·55	56
1·0	1·0	0
1·5	0·78	−39
2·0	0·56	−56·5

Comment

The pole-zero diagram, in affording a phasor interpretation of response solely in terms of the fixed natural frequencies of the network and the variable frequency of the excitation, highlights that the response of a network to a given excitation is, except for a scale factor, fundamentally a function only of the differences between the excitation frequency and the natural frequencies or poles and zeros.

The common type of phasor diagram for ac circuits is for a fixed circuit and frequency. As frequency is the only variable in the complex-frequency plane, the pole-zero diagram is superior for assessing frequency dependence, not only on excitation, but also on the natural frequencies which could be allocated by trial without the initial involvement of circuit elements. It is in this sense that it is the cornerstone of network synthesis, in which a desired response is expressed as a function of poles and zeros, and a network is then realised which conforms to the function.

The complex-frequency plane is important also for its generality, for it applies also to an excitation frequency of form $s = \sigma + j\omega$ which represents an excitation of form $e(t) = Ee^{\sigma t} \cdot e^{j\omega t}$. This is its role in transient response, notably as found with the Laplace transform, through which the time variable in a differential equation is temporarily transformed into the complex-frequency variable s in a commutative algebraic equation that yields the complete solution as a function of time after manipulation and inverse transformation. The present illustration is really a special case in which $s = 0 + j\omega$ and the state is purely sinusoidal and steady: the response in the time-domain for $R = 1\,\Omega$ when $\omega = 0.5$ rad/s and $s = j0.5$ is $i(t) = 0.55 \sin(\omega t + 56\pi/180)$ or $i(t) = 0.55e^{j(\omega t + 56\pi/180)}$.

ILLUSTRATION 2.12

What is the Laplace transform, and what is its role in network and system theory?

Interpretation

The Laplace transform,

$$F(s) = \int_0^\infty f(t)e^{-st}\,dt \tag{1}$$

is a relationship through which a function having time t as the variable may be transformed into one having a complex frequency $s = \sigma + j\omega$

as the variable; and in consequence, and of special moment, through which the operations of differentiation and integration are changed, respectively, into multiplication and division by s. Thus may a differential equation in the time domain (in which both excitation and response are functions of time) be transformed temporarily into an algebraic equation in the frequency domain, for manipulation freely into an expression of the solution as an explicit function of the frequency variable s. This solution may be reinstated in the time domain through the inverse transform,

$$\mathcal{L}^{-1}F(s) = f(t) = \frac{1}{2\pi j} \int_{\sigma-j\infty}^{\sigma+j\infty} F(s)\, e^{st}\, ds \tag{2}$$

In practice, neither equation (1) nor equation (2) requires evaluation; for the transforms have been tabulated for a very wide range of functions. Simple examples are:

$$f(t) = U(t) = 1_{(t\,>\,0)}, \quad F(s) = 1/s \tag{3a}$$
$$f(t) = t^0_{(t\,>\,0)} = 1, \quad F(s) = 1/s \tag{3b}$$
$$f(t) = e^{\alpha t}, \quad F(s) = 1/(s-\alpha) \tag{3c}$$
$$f(t) = \sin \omega t, \quad F(s) = \omega/(s^2+\omega^2) \tag{3d}$$
$$f(t) = \cos \omega t, \quad F(s) = s/(s^2+\omega^2) \tag{3e}$$

while, in the presence of an initial value $F(0)$, the Laplace transform of the derivative of $f(t)$ is

$$\mathcal{L}Df(t) = sF(s) - F(0)$$

Now consider the solution of Illustration 2.2 by temporary transformation into the s-domain. In Figure 2.1 there is an initial current $I(0) = 10/10 = 1$ A at the moment S is closed. Thereafter the circuit resistance is reduced from $10\,\Omega$ to $7\cdot5\,\Omega$, and the transformed differential equation is the algebraic equation,

$$L[sI(s)-1]+7\cdot5I(s) = E/s = 10/s$$

where $I(s)$ replaces $i(t)$ while the constant emf E is replaced by its transform E/s in accordance with eqn. 3(b). Manipulating and separating into partial fractions gives

$$I(s) = \frac{1+s}{s(s+\frac{3}{4})} = \frac{1}{3}\left[\frac{4}{s} - \frac{1}{s+\frac{3}{4}}\right]$$

But by inspection of equations (3), the terms in the brackets are recognisable as the transforms of $f_1(t) = 4$ and $f_2(t) = e^{-3t/4}$. The solution $I(s)$ transformed back into the time domain is therefore

$$i(t) = \mathcal{L}^{-1}I(s) = \tfrac{4}{3}[1 - \tfrac{1}{4}e^{-3t/4}]$$

as in Illustration 2.2.

Comment

In this instance, solution by Laplace transformation is simpler (except that partial fractions are involved) than direct solution in the time domain, as in Illustration 2.2; for the initial condition is included in the working, and the result is the complete solution. But in more complicated cases, the advantage of unrestricted algebraic manipulation may be offset by the difficulty of resolving $F(s)$ into a form that readily reveals the inverse transforms of its parts.

By contrast, the transient part of the response of a linear system, being the solution of the homogeneous or excitation-free differential equation, is always quotable in the form $A_n e^{s_n t} + A_{n-1} e^{s_{n-1} t} + \ldots$, where the natural frequencies are calculable from the system or circuit elements, and the constants are governed by initial conditions. Such constants may, however, be tedious to find when numerous. Moreover, finding the steady-state component (particular integral) from the differential equation in the time domain is intractable without a computer, except for a restricted range of excitation forms (such as E, Ee^{st}, $E \sin \omega t$ and a few others).

On the other hand, Laplace transformation yields the complete solution, in which is included the steady-state component, for a very wide range of excitations that might otherwise be intractable. Nevertheless, transformation does not necessarily afford a quick or easy solution to such cases. A good example is that of a sawtooth excitation (as used for beam deflection in an oscilloscope, or scanning in television) applied to a capacitor in series with a resistor. Because the transform for such an excitation, of recurrence period T, has the form

$$E(s) = \frac{E}{Ts^2(1-e^{-Ts})} [1-(Ts+1)e^{-Ts}]$$

$F(s)$ becomes involved, in spite of the simplicity of the circuit, and extraction of the inverse transform for $i(t)$ or $v_R(t)$ requires several printed pages (see, for example, D. K. Cheng: 'Analysis of Linear Systems', p. 209–211).

The modern computer is able to solve suitably programmed differential equations directly. In fact, its incremental mode of operation is akin to the basic concept of calculus, and in negligible time it is able to calculate the transient response of a system or network of such complexity and under such conditions that calculation by hand would be impossible. In this respect, the transform method has lost some of its importance as a technique for transient solution, although a computer can be programmed in terms of it.

Perhaps the most important feature of the Laplace transform is that equation-forms for a given network in terms of the complex variable s are independent of the kind of excitation, if the excitation and response are attributed in general the forms $E(s)$ and $I(s)$. This implies that the forms are universal, and adaptable to any particular excitation function out of the wide variety that is Laplace-transformable. Equivalently, there is no need to distinguish between the transient and the steady state, or between sinusoidal and non-sinusoidal excitations. Thus the impedance and admittance concepts can be generalised in the forms

$$Z(s) = E(s)/I(s), \quad Y(s) = I(s)/E(s)$$

For the steady sinusoidal state, in which an excitation of rotating phasor form $|\hat{E}|e^{j\omega t}$ forces the response into a similar pattern $|\hat{I}|e^{j(\omega t+\phi)}$ after the transitory interval in which the network or system asserts its natural modes against the overwhelming continuous force of the excitation, the immittance relations become.

$$Z(s) = \frac{E(j\omega)}{I(j\omega)} = Z(j\omega), \quad Y(s) = \frac{I(j\omega)}{E(j\omega)} = Y(j\omega)$$

so that s, then a variable on the $j\omega$ axis of the complex-frequency plane, has the particular form $s = j\omega$.

ILLUSTRATION 2.13

Figure 2.11 is a low-pass filter scaled to a cut-off frequency $\omega_c = 1$ rad/s and to terminating resistances of $1\,\Omega$.

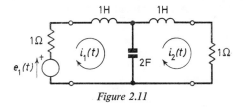

Figure 2.11

(a) Find the transfer admittance in the generalised form $Y(s) = I_2(s)/E_1(s)$.

(b) Find $i_2(t)$ when $e_1(t) = U(t)E_1$.

(c) Find an expression for the steady-state response $I_2(j\omega)$ to a sinusoidal excitation $E_1(j\omega)$.

Interpretation

(a) Let the arbitrary excitation $e_1(t)$ be Laplace transformed. Then functionally it becomes $E_1(s)$, while $i_1(t)$ and $i_2(t)$ become $I_1(s)$ and $I_2(s)$. In the s-domain the mesh impedances can be written directly in the form $Z(s) = R+sL+1/sC$, so that on substituting values,

$$(1+s+1/2s)\,I_1(s)-I_2(s)/2s = E_1(s)$$

$$(1+s+1/2s)\,I_2(s)-I_1(s)/2s = 0$$

whence

$$Y(s) = \frac{I_2(s)}{E_1(s)} = \frac{1}{2(1+s)(1+s+s^2)} \tag{1}$$

(b) When $e_1(t) = U(t)E_1$, $E_1(s) = \mathcal{L}e_1(t) = E_1/s$
Then,

$$I_2(s) = E_1(s)\,Y(s) = \frac{E_1}{2}\left[\frac{1}{s(1+s)(1+s+s^2)}\right] = \frac{1}{2}E_1 F(s) \tag{2}$$

The denominator of $F(s)$ can be factorised further by setting it to zero and finding the roots, giving

$$F(s) = \frac{1}{s(s-p_1)(s-p_2)(s-p_3)} \tag{3}$$

where

$$p_1 = -1, \quad p_2 = -\frac{1}{2}+j\frac{\sqrt{3}}{2}, \quad p_3 = -\frac{1}{2}-j\frac{\sqrt{3}}{2}$$

are the natural frequencies or poles of $F(s)$.

Equation (3) expands into partial fractions in the form

$$F(s) = \left[\frac{K}{s}+\frac{K_1}{s-p_1}+\frac{K_2}{s-p_2}+\frac{K_3}{s-p_3}\right] \tag{4}$$

Consider now the form

$$f(s) = \left[\frac{K_1}{s-p_1}+\frac{K_2}{s-p_2}+\dots+\frac{K_k}{s-p_k}+\dots+\frac{K_n}{s-p_n}\right]$$

If both sides are multiplied by $s-p_k$ and the condition $s = p_k$ is imposed after this is done, all terms vanish except K_k. Thus,

$$K_k = [(s-p_k)\,f(s)]_{s\,=\,p_k}$$

Applying this principle to equation (4), where $f(s) = F(s)$ in equation (2),

$$K = [sF(s)]_{s\,=\,0} = 1$$
$$K_1 = [(s-p_1)F(s)]_{s\,=\,p_1} = -1$$

and by similar procedure, $K_2 = j/\sqrt{3}$ and $K_3 = -j/\sqrt{3}$. Then,

$$i_2(t) = \frac{E_1}{2} \mathcal{L}^{-1} \left[\frac{1}{s} - \frac{1}{s+1} + \frac{j}{\sqrt{3}} \cdot \frac{1}{s+(\frac{1}{2}-j\sqrt{3}/2)} \right.$$

$$\left. - \frac{j}{\sqrt{3}} \cdot \frac{1}{s+(\frac{1}{2}+j\sqrt{3}/2)} \right]$$

$$= \frac{E}{2} \left[1 - e^{-t} + \frac{j}{\sqrt{3}} e^{-(1/2-j\sqrt{3}/2)t} - \frac{j}{\sqrt{3}} e^{-(1/2+j\sqrt{3}/2)t} \right]$$

$$= \frac{E}{2} \left[1 - e^{-t} - \frac{2}{\sqrt{3}} e^{-t/2} \sin \frac{\sqrt{3}}{2} t \right]$$

The inverse transforms have been recognised from equations (3), Illustration 2.12.

(c) The steady-state response to $E_1(j\omega)$ is given immediately on putting $s = j\omega$ in equation (1). This gives

$$I_2(j\omega) = E_1(j\omega) Y(j\omega) = \frac{E_1(j\omega)}{2(1+j\omega)(1+j\omega-\omega^2)}$$

$$= \frac{E_1(j\omega)}{2[(1-2\omega^2)-j(\omega^3-2\omega)]}$$

and

$$|I_2(j\omega)| = \frac{1}{2} \cdot \frac{|E_1(j\omega)|}{\sqrt{[(1-2\omega^2)^2+(\omega^3-2\omega)^2]}}$$

Comment

This analysis demonstrates the generality afforded by Laplace transformation, and the adaptability of the common $F(s)$ to two different excitations.

ILLUSTRATION 2.14

An emf $e(t) = \hat{E} \sin \omega t$ is applied to an inductor L in series with a resistor R at an instant $t = 0$ when the emf is passing through zero. Obtain an expression for the current at $t > 0$ directly in the time domain, assuming that commonplace ac theory may be invoked. Compare this approach with Laplace transformation when the form $\mathcal{L} \sin \omega t = \omega/s^2+\omega^2)$ is used.

How are the approaches and solutions modified when the emf is applied at an instant $t = t_1$, where $t_1 > 0$ and the emf is $e(t_1)$ at that instant?

Interpretation

Let $i = i(t)$. Then in the time domain,

$$Ri + L\,Di = \hat{E} \sin \omega t \qquad (t > 0) \tag{1}$$

The homogeneous equation $Ri + L\,Di = 0$ is completely independent of excitation, and its quotable solution is

$$i = i_{tr} = Ae^{st}, \quad \text{where} \quad s = -R/L \tag{2}$$

The steady-state solution is also quotable from simple ac theory as

$$i_{ss} = \hat{E} \sin (\omega t - \phi)/|Z| \tag{3}$$

when $|Z| = \surd(R^2 + \omega^2 L^2)$ and $\phi = \tan^{-1}(\omega L/R)$

At $t = 0$, $i(0) = i_{tr}(0) + i_{ss}(0) = 0$. Then, putting $t = 0$ in equations (2) and (3),

$$A = -i_{ss}(0) = -\hat{E} \sin (-\phi)/|Z| = E \sin \phi/|Z| \tag{4}$$

and the complete solution to equation (1) is therefore

$$i(t) = \hat{E}[\sin \phi e^{-Rt/L} + \sin (\omega t - \phi)]/|Z| \tag{5}$$

Solution by Laplace transformation with the given excitation transform requires resolution of

$$I(s) = \hat{E}\omega/(s^2 + \omega^2)(sL + R) \tag{6}$$

into partial fractions affording comparison with known transforms, such as those in equations (3), Illustration 2.12. Equation (6) may be resolved into the form

$$I(s) = \frac{\hat{E}\omega L}{|Z|^2} \left[\frac{1}{s + R/L} - \frac{s}{s^2 + \omega^2} + \frac{R/L}{s^2 + \omega^2} \right]$$

$$= \frac{\hat{E} \sin \phi}{|Z|} \left[\frac{1}{s + R/L} - \frac{s}{s^2 + \omega^2} + \frac{R}{\omega L} \cdot \frac{\omega}{s^2 + \omega^2} \right] \tag{7}$$

where $|Z|$ and ϕ are defined as before. Then,

$$i(t) = \mathcal{L}^{-1}I(s) = \frac{\hat{E} \sin \phi}{|Z|} \left[e^{-Rt/L} - \cos \omega t + \frac{R}{\omega L} \cdot \sin \omega t \right]$$

$$= \hat{E}[\sin \phi e^{-Rt/L} + \sin (\omega t - \phi)]/|Z| \tag{8}$$

which corresponds to equation (5).

When the emf is applied at an instant $t_1 > 0$, its instantaneous value is $e(t_1) = \hat{E} \sin \omega t_1$, and the relation

$$i(t_1) = i_{tr}(t_1) + i_{ss}(t_1) = 0 \quad \text{gives}$$
$$i_{tr}(t_1) = -i_{ss}(t_1) = -\hat{E} \sin (\omega t_1 - \phi)/|Z|$$

Equations (2) and (3) now give the complete solution

$$i(t)_{(t > t_1)} = \hat{E}[-\sin (\omega t_1 - \phi)e^{-R(t - t_1)/L} + \sin (\omega t - \phi)]/|Z| \quad (9)$$

For a Laplace transform approach the excitation may be symbolised by a delayed step, in the form $U(t - t_1)\hat{E} \sin \omega t$. But,

$$\mathcal{L}U(t - t_1) \sin \omega t = \int_{t_1}^{\infty} \sin \omega t e^{-st} \, dt$$

$$= e^{-t_1 s} \frac{\omega \cos \omega t_1 + s \sin \omega t_1}{s^2 + \omega^2} \quad (10)$$

Though more involved, this merely replaces t by $t - t_1$ in the transitory component of the time-domain solution, as in equation (9). It does not exemplify the *shifting theorem*, which applies to the form $U(t - t_1)F(t - t_1)$. For this the Laplace transform is simply $e^{-t_1 s}\mathcal{L}F(t)$; and therefore

$$\mathcal{L}[U(t - t_1) \sin \omega(t - t_1) = e^{-t_1 s}\omega/(s^2 + \omega^2)$$

which is simpler than equation (10).

Comment

This illustrates the relative simplicity of solution from the time-domain differential equation when the excitation has a well-known steady-state response (particular integral) and the homogeneous equation is of low order. It emphasises again that the transient component of response depends, not on the form of excitation, but only on initial conditions. Note that a shorter Laplace transform solution would be afforded by a more comprehensive list of transforms.

ILLUSTRATION 2.15

A network has three nodes 1, 2, 3. Between nodes 1 and 3 are three parallel branches comprising a unidirectional time-invarient emf E in series with a resistor R, a resistor R_1 in series with a switch S, and a capacitor C; between nodes 1 and 2 is a resistor R_2; and between nodes 2 and 3 is an inductor L whose resistance is negligible.

The circuit has been in a steady state for a long time with the switch S closed. At an instant $t = 0$ the switch S is opened. Find a numerical expression for the voltage $v_2(t)$ across L at an instant $t > 0$ when $R_2 = R_1 = R = 100\,\Omega$, $C = 1\,\mu F$, $L = 10\,mH$, and $E = 100\,V$.

Interpretation

The topological arrangement and variable to be found suggest nodal-voltage analysis. Accordingly, let $G = 1/R$, $G_1 = 1/R_1$ and $G_2 = 1/R_2$. Transforming E, R into a current source $I = GE$ in parallel with G, the elements between nodes 1 and 3 for $t < 0$ are I, G, G_1 and C; while for $t > 0$ they are I, G and C. Let node 3 be the datum, and let v_1 and v_2, both assumed positive, denote the node-datum potentials of nodes 1 and 2. The initial steady state impinging on $t = 0$ determines an initial capacitor voltage $V_1(0) = I/(G+G_1+G_2)$ and an initial inductor current $I_L(0) = IG_2/(G+G_1+G_2)$. Then, noting that in the presence of an initial value, $\mathcal{L}Df(t)$ has the form $sF(s)-F(0)$, the Laplace-transformed time-domain equations for $t > 0$ can be written by inspection as

$$(G+G_2)V_1(s)+C[sV_1(s)-V_1(0)]-G_2V_2(s) = I/s \tag{1}$$

$$-G_2V_1(s)+I_L(0)/s+(G_2+1/sL)V_2(s) = 0 \tag{2}$$

where the current source I is assumed to flow from node 3 to node 1. In equation (2),

$$\frac{V_2(s)}{sL} = \mathcal{L}\left[\frac{1}{L}\int_0^t v_2(t)\,dt\right]$$

Putting $G_2 = G_1 = G$ and manipulating equations (1) and (2) gives

$$[Cs^2+(G+C/LG)s+2/L]V_2(s) = I-2I_L(0)+sC[V_1(0)-I_L(0)/G] \tag{3}$$

whence

$$V_2(s) = \frac{a+bs}{\alpha s^2+\beta s+\gamma} \tag{4}$$

where, substituting numerical values, $I = 1$ A, $I_L(0) = I/3 = 1/3$ A, $V_1(0) = I/3G = 10^2/3$ V, and

$$a = I - 2I_L(0) = 1/3; \quad b = C[V_1(0) - I_L(0)/G] = 0;$$
$$\alpha = C = 10^{-6}; \quad \beta = G + C/LG = 2 \times 10^{-2}; \quad \gamma = 2/L = 2 \times 10^2$$

Thus,

$$V_2(s) = \frac{10^6/3}{s^2 + 2 \times 10^4 s + 2 \times 10^8}$$

$$= \frac{10^2}{3} \cdot \frac{10^4}{(s + 10^4)^2 + (10^4)^2} \tag{5}$$

which, by reference to a table of Laplace transforms, gives

$$v_2(t) = \mathcal{L}^{-1} V_2(s) = 33 \cdot 3 e^{-10^4 t} \sin 10^4 t \tag{6}$$

Comment

This case shows how a circuit of mixed elements may involve initial values both for charges of the form $CV(0)$ in capacitors and for flux-linkages of the form $LI(0)$ in inductors.

The denominator of equation (4) contains the poles or natural frequencies in $v_2(t)$, and is identical with the characteristic equation that might have been found from the homogeneous equation in the time domain; but wheras this would have required separate determination of arbitrary constants from the initial conditions, these are carried in the Laplace-transformed equations, which yield the complete solution through one algebraic process and inversion.

CHAPTER 3

Simplifying procedures, theorems and equivalences

INTRODUCTION

This chapter is devoted to the illustration of procedures that may ease or even obviate formal analysis.

Scaling in respect of magnitudes and frequency is the first consideration, for it usually makes numerical values more tractable. It is fundamental to synthesis procedures, as the polynomial forms in which prescribed immittance functions are represented should have their real, positive coefficients as integers of the lowest values, to facilitate manipulation and the realisation of a network. This requirement is satisfied by the choice of 1 Ω as the reference for impedance magnitude and $\omega = 1$ rad/s as the reference for frequency (see, for example, Illustrations 4.15 and 4.16).

The principal theorems considered in the context of simplification are the Thevenin–Helmholtz theorem and its dual, Norton's theorem. They are of paramount importance in justifying replacement of the whole of an activated network, in respect of its output port, by a single immittance and a single source (which may need to be complex if there are several sources of different frequencies within the network). These theorems act respectively as mesh and node reducing artifices, and are very useful for the replacement of the whole of a relatively simple network (e.g. a T or π with input generator), or part of a complicated one (to which they can be applied in an iterative way), provided the parameters of the equivalent simple generator are obvious or easy to calculate.

The equivalences considered are the star-delta transformation, lattice network transformations, and alternative representations for a mutual inductor. Each case reflects a change in the topology of the network, so that an arrangement that is awkward to analyse might be changed into a more convenient externally equivalent one. The impact is striking in the case of the lattice network, when it is transformed easily with Bartlett's bisection theorem into an equivalent balanced T structure; and again in the case of a mutual inductor, which is ill-adapted to nodal voltage analysis in the form of a transformer, but well adapted to it when replaced by an equivalent π network of self-inductors (see Illustrations 3.20 and 3.21).

The digital computer has revolutionised network calculations, and is able to solve nodal-voltage and loop-current equations for very complicated networks. This means that the simplifications considered here may have little relevance to computer programs: there is little point in preliminary simplification with the Thevenin–Helmholtz theorem or with equivalences, if a program can be compiled for solution directly in terms of Kirchhoff's laws. Nevertheless, the engineer is always faced with on-the-spot calculations and estimates on moderate networks not justifying immediate use of a computer; and moreover, the full potential of the computer can only be realised, perhaps, from a good understanding of network theory.

ILLUSTRATION 3.1

Scale the network and sources in Figure 3.1 so that when the load R is reduced to $1\,\Omega$, the currents I_1 and I_2 are the same as in the actual network. Show that the powers P_1 and P_2 entering and leaving the network are both scaled by the same factor as the branch resistances, but that the ratio P_1/P_2 is unaltered by scaling.

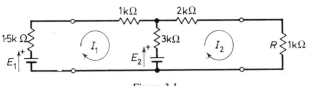

Figure 3.1

Interpretation

The scaled network is shown in Figure 3.2. Its validity in preserving I_1 and I_2 lies simply in the validity of scaling the coefficients in a set of linear equations by the same factor on each side. The equations for the

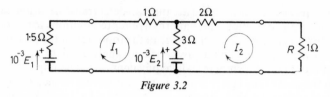

Figure 3.2

actual network,

$$5500I_1 - 3000I_2 = E_1 - E_2$$
$$-3000I_1 + 6000I_2 = E_2$$

are unchanged when divided through by 10^3 to give

$$5 \cdot 5I_1 - 3I_2 = 10^{-3}(E_1 - E_2)$$
$$-3I_1 + 6I_2 = 10^{-3}E_2$$

which are the equations for the scaled network. The currents in the scaled network have thus the same values as in the actual, provided the emf sources are also divided by the resistance-scaling factor.

The power entering at the actual network terminals is

$$P_{1(a)} = (E_1 - 1 \cdot 5 \times 10^3 I_1)I_1$$

while that entering the scaled network is

$$P_{1(s)} = (10^{-3}E_1 - 1 \cdot 5I_1)I_1 = 10^{-3}(E_1 - 1 \cdot 5 \times 10^3 I_1)I_1$$

so that

$$P_{1(s)} = 10^{-3}P_{1(a)}$$

Similarly, $P_{2(a)} = 10^3 I_2^2$, $P_{2(s)} = 1 \times I_2^2$ and $P_{2(s)} = 10^{-3}P_{2(a)}$. But from these results

$$\frac{P_{2(a)}}{P_{1(a)}} = \frac{10^3 P_{2(s)}}{10^3 P_{1(s)}} = \frac{P_{2(a)}}{P_{1(s)}}$$

and therefore the input-output power ratio is independent of the scaling.

Comment

(1) That the input-output power ratio is independent of scaling is important in connection with the transfer function of a two-port network. In the case of a reactive network, the phase-change is also independent of scaling.

(2) While the circuit as a whole is linear, the two-port T-network bounded by the terminals is an active one by virtue of the presence of

E_2, and it behaves externally as a non-linear network. This is evident from the input resistance when the source E_1, R_1 is removed. Let a voltage V_1 be applied to the input terminals of the scaled network. Then

$$R_{in} = \frac{V_1}{I_1} = \frac{5V_1}{2V_1 - 10^{-3}E_2}\ \Omega$$

This is not constant, as for a linear passive network, but is voltage dependent. When E_2 is reduced to zero, the network becomes passive and the input resistance assumes the constant value

$$R_{in} = \frac{V_1}{I_1} = \frac{5}{2}\ \Omega$$

which is apparent also from the series-parallel combination of resistors.

ILLUSTRATION 3.2

The emf E_2 in Figure 3.1 is replaced by a current source I_{s2} in parallel with the shunt branch, and acting towards its junction with the series branches. Scale the network to a $1 - \Omega$ base, and scale the sources so that the actual currents are preserved.

Interpretation

The scaled network is shown in Figure 3.3.

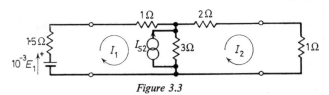

Figure 3.3

The equations for the actual network may be written directly in the forms

$$5500I_1 - 3000(I_2 - I_{s2}) = E_1$$
$$-3000(I_1 + I_{s2}) + 6000I_2 = 0$$

in which I_{s2} is treated like an additional mesh current, or in the forms

$$5500I_1 - 3000I_2 = E_1 - 3000I_{s2}$$
$$-3000I_1 + 6000I_2 = 3000I_{s2}$$

in which I_{s2} and the $3k\,\Omega$ shunt branch have been transformed into an equivalent emf generator. Either form leads, on being divided through by 10^3, to the scaled circuit of Figure 3.3.

Comment

While the emf is scaled down by 10^3, the current source is unchanged. This is not only consistent with the arithmetic, but also with the requirement that the currents are to be unchanged: this must include the current source itself for correct proportionality.

ILLUSTRATION 3.3

The scaled circuit of Figure 3.4 is the dual of Figure 3.3. Confirm this by formulating the nodal-voltage equations, and scale the network and sources to preserve the node-datum voltages when the load conductance is reduced to 1 m S.

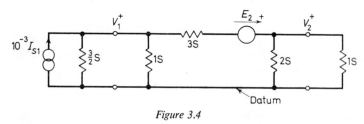

Figure 3.4

Interpretation

The node-datum voltage equations may be written directly in the forms

$$5 \cdot 5V_1 - 3(V_2 - E_2) = 10^{-3}I_{s1}$$
$$-3(V_1 + E_2) + 6V_2 = 0$$

in which E_2 is treated as an extra node-voltage subtracting from V_2 or adding to V_1, or in the forms

$$5 \cdot 5V_1 - 3V_2 = 10^{-3}I_{s1} - 3E_2$$
$$-3V_1 + 6V_2 = 3E_2$$

in which a transformation of the series branch into an equivalent current generator has been effected. The form of these equations is identical with that for Figure 3.3.

Scaling Figure 3.4 for a load conductance of 1 m S requires division of the equations by 10^3. The current source thus becomes $10^{-6}I_{s1}$, the emf source E_2 is unchanged, while the network branches have resistance values, from left to right, of $\frac{2}{3}$ kΩ ($\frac{3}{2}$ m S), 1 kΩ (1 m S), $\frac{1}{3}$ kΩ (3 m S), $\frac{1}{2}$ kΩ (2 m S) and 1 kΩ (1 m S).

Comment

The duality of the networks is implicit in the identity of equation forms when the parameters are systematically interchanged: exchanging conductance for resistance, node voltages for mesh currents, and current sources for emf sources and vice-versa transforms the equations for Figure 3.4 into those for Figure 3.3.

ILLUSTRATION 3.4

A circuit comprises a resistor R_1, an inductor L having a series resistance r, and a capacitor C, all connected in series.

1. If the circuit has a steady-state impedance $Z(j\omega_1) = R + jX(\omega_1) = |Z(j\omega_1)| \; \underline{/\phi}$ at an angular frequency ω_1, by what factors must L and C be multiplied in order to preserve this impedance at a new angular frequency $\omega_2 = \omega_1/\beta$?

2. If R_1 is changed to $R_2 = R_1/\alpha$, by what factors must r, L, and C be multiplied,

(a) to preserve ϕ at ω_1?
(b) to preserve ϕ at ω_2?

3. In what respect is Z changed under conditions 2(a) and 2(b), and what simple conclusions may be drawn in respect of scaling of complex impedances?

Interpretation

1. L and C must both be multiplied by β; for then,

$$X(\omega_2) = \frac{\omega_1}{\beta}(L\beta) - \frac{\beta}{\omega_1(C\beta)} = X(\omega_1)$$

2. (a) Both r and L must be divided by α, but C must be multiplied by α; for then,

$$Z(j\omega_1) = \frac{R_1}{\alpha} + \frac{r}{\alpha} + j\left(\omega_1 \cdot \frac{L}{\alpha} - \frac{1}{\omega_1 C\alpha}\right)$$

$$= \frac{1}{\alpha}[R + jX(\omega_1)]$$

and
$$\phi = \tan^{-1} X(\omega_1)/R \text{ is unchanged.}$$

(b) To preserve ϕ at ω_2, L and C must also be multiplied by β; for then,

$$Z(j\omega_2) = \frac{R_1}{\alpha} + \frac{r}{\alpha} + j\left[\frac{\omega_1}{\beta} \cdot L \cdot \frac{\beta}{\alpha} - \frac{\beta}{\omega_1 C \alpha \beta}\right]$$

$$= \frac{1}{\alpha}[R + jX(\omega_1)] \quad \text{as before.}$$

3. $Z(j\omega)$ is altered in magnitude only, by the factor α.

It is to be concluded that when the impedance of a network branch is complex, and this impedance is to be preserved in argument at a new frequency but scaled in magnitude, the elements R, L, and C comprising the branch must be modified to have the scaled-branch values

$$R_{(s)} = \frac{R}{\alpha}, \quad L_{(s)} = L\left(\frac{\beta}{\alpha}\right), \quad C_{(s)} = C(\alpha\beta)$$

when

α is the magnitude-scaling factor

β is the frequency-scaling factor

α and β can be expressed in the forms

$$\alpha = R/R_{(n)}$$

where $R_{(n)}$ is the new level of reference for magnitude, and

$$\beta = \omega/\omega_{(n)}$$

when $\omega_{(n)}$ is the new reference for angular frequency.

Comment

$R_{(n)}$ is commonly taken as $1\,\Omega$, and $\omega_{(n)}$ as 1 rad/s. Networks referred to these levels of impedance magnitude and angular frequency are sometimes called normalised networks (other norms could, however, be used).

ILLUSTRATION 3.5

A sinusoidal emf $E(j\omega)$ is applied to a series circuit consisting of an inductor L of series resistance R, in series with a capacitor C. Show that the response $I(j\omega)$ may be expressed as a function of the ratio ω_r/ω, where ω_r is the resonant frequency, and that this response is independent of scaling provided the ratio $\omega_r L/R$ is constant.

Interpretation

The angular resonant frequency for a series LCR circuit, given by $\omega_r^2 = 1/LC$, is that frequency at which the current is a maximum under constant conditions of excitation, and the impedance is a minimum, being equal to the series resistance R.

The response $I(j\omega)$ is

$$I(j\omega) = E(j\omega) \cdot Y(j\omega)$$

$$= \frac{E(j\omega)}{R + j\left(\omega L - \dfrac{1}{\omega C}\right)}$$

$$= \frac{E(j\omega)/R}{1 + j\left(\omega \cdot \dfrac{L}{R} - \dfrac{1}{\omega} \cdot \dfrac{1}{CR}\right)}$$

$$= \frac{E(j\omega)/R}{1 + j\left(\omega \cdot \dfrac{L}{R} - \dfrac{\omega_r^2}{\omega} \cdot \dfrac{L}{R}\right)}$$

Let $\omega_r L/R = Q_r$. Then $L/R = Q_r/\omega_r$
and

$$I(j\omega) = \frac{E(j\omega)/R}{1 + jQ_r\left[\dfrac{\omega}{\omega_r} - \dfrac{\omega_r}{\omega}\right]}$$

or,

$$|I(j\omega)| = \frac{I_r}{\left\{1 + Q_r^2\left[\dfrac{\omega}{\omega_r} - \dfrac{\omega_r}{\omega}\right]^2\right\}^{1/2}}$$

and

$$\arg I(j\omega) = -\tan^{-1} Q\left[\dfrac{\omega}{\omega_r} - \dfrac{\omega_r}{\omega}\right]$$

Since, for constant excitation and Q-factor, both magnitude and phase are functions of the frequency ratio only, neither the magnitude-frequency nor phase-frequency curve is altered in shape by scaling, provided the parameters L, R, and C are scaled to values

$$L_{(s)} = L \cdot \left(\frac{\beta}{\alpha}\right) \quad \text{and} \quad C_{(s)} = C(\alpha\beta)$$

where

$$\alpha = R/R_{(s)} \quad \text{and} \quad \beta = \omega_r/\omega_{r(s}$$

Comment

For the common case of scaling to an angular reference frequency of 1 rad/s and a resistance level of 1 Ω, $\beta = \omega_r$ and $\alpha = R$ so that

$$L_{(s)} = L\left(\frac{\omega_r}{R}\right) \quad \text{and} \quad C_{(s)} = C(R\omega_r)$$

ILLUSTRATION 3.6

A symmetrical T-section filter, scaled to terminating resistors $R_t = 1\,\Omega$ and a cut-off frequency $f_c = 1/2\pi$ Hz, has inductors $L_a = 0\cdot6$ H for its series arms, and a shunt arm comprising an inductor $L_b = 0\cdot534$ H in series with a capacitor $C_b = 1\cdot2$ F. Determine the filter elements L_a, L_b and C_b for a practical application in which $f_c = 5\times10^5/2\pi$ Hz and $R_t = 75\,\Omega$.

Obtain an expression for the transfer function $I_2(s)/E_1(s)$ of the practical filter, where $E_1(s)$ is the emf of a 75-Ω generator connected to its input terminals, $I_2(s)$ is the resultant current in the 75-Ω load resistor, and s may be regarded as a contraction for the steady sinusoidal-state operator $j\omega$. For what values of s is $I_2(s)$ zero?

Interpretation

(a) For $f_c = 5\times10^5/2\pi$ Hz or $\omega_c = 5\times10^5$ and $R_t = 75\,\Omega$,

$$L_a = \frac{0\cdot6\times75\times10^6}{5\times10^5} = 90\ \mu\text{H}$$

$$L_b = \frac{0\cdot534\times75\times10^6}{5\times10^5} = 80\cdot1\ \mu\text{H}$$

$$C_b = \frac{1\cdot2\times10^6}{5\times10^5\times75} = 0\cdot032\ \mu\text{F}$$

(b) The transfer function for the actual filter has the same form as that for the scaled filter, which is much easier to calculate. Denoting clockwise currents in the first and second meshes as $I_1(s)$ and $I_2(s)$, and writing sL and $1/sC$ in place of $j\omega L$ and $1/j\omega C$, the mesh equations are

$$Z_{11}(s)I_1(s)+Z_{12}(s)I_2(s) = E_1(s)$$
$$Z_{21}(s)I_1(s)+Z_{22}(s)I_2(s) = 0$$

where

$$Z_{11}(s) = Z_{22}(s) = 1+1\cdot134s+1/1\cdot2s$$
$$Z_{12}(s) = Z_{21}(s) = -(0\cdot534s+1/1\cdot2s)$$

Solving for $I_2(s)$ gives

$$\frac{I_2(s)}{E_1(s)} = \frac{1+0.64s^2}{2(1+0.6s)(1+0.6s+s^2)} \tag{1}$$

or, as the ratio of two polynomials in s,

$$\frac{I_2(s)}{E_1(s)} = \frac{1+0.64s^2}{2+2.4s+2.72s^2+1.2s^3} \tag{2}$$

Equation (1) factorises further, and the factors may be put in the form

$$\frac{I_2(s)}{E_1(s)} = \frac{1}{2}\frac{(s-z_1)(s-z_2)}{(s-p_1)(s-p_2)(s-p_3)} \tag{3}$$

where

$$z_1 = j1.25, \qquad z_2 = -j1.25$$
$$p_1 = -1.67, \qquad p_2 = -0.3+j0.955, \qquad p_3 = -0.3-j0.955$$

$I_2(s) = 0$ when $s-z_1 = 0$ or $s-z_2 = 0$, which is so when

$$s = z_1 = j1.25 = j\omega_z \quad \text{where} \quad \omega_z = 1.25, \quad \text{or}$$
$$s = z_2 = -j1.25 = -j\omega_z$$

Comment

The roots z_1, z_2 and p_1, p_2, p_3, in the factored form, equation (3), are called, respectively, the zeros and poles of the function. They represent the values of the variable s for which the function becomes zero and infinity. In the case illustrated, p_1 is negative real while p_2 and p_3 are complex conjugates (they are, in fact, complex natural frequencies for the system). The transfer function can therefore never become infinite for purely sinusoidal excitation, which coresponds to $s = j\omega$ and wholly real frequency ω, although it can become zero at real frequency, since the zeros are wholly imaginary and can be cancelled (as shown) when $s = j\omega_z$. The significance of poles and zeros and the complex-frequency concept is illustrated more fully in Chapter 2.

ILLUSTRATION 3.7

The scaled tuned transformer shown in Figure 3.5 is required to match a generator of internal resistance $R_1 = 10\,\Omega$ connected across terminals 1–2 to a load of resistance $R_2 = 1\,\Omega$ connected across

8

terminals 3–4, and at an angular frequency of 1 rad/s. Calculate the values required for the coupling coefficient k and capacitance of C_1.

What are the practical component values when $R_1 = 10$ kΩ, $R_2 = 1$ kΩ, and the frequency is 1 MHz?

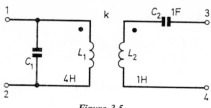

Figure 3.5

Interpretation

Let Z_{22} denote the contour impedance of the mesh formed by closing terminals 3–4 through a load. Then an impedance

$$Z = \cdot \frac{\omega^2 M^2}{Z_{22}} = \frac{\omega^2 k^2 L_1 L_2}{Z_{22}}$$

is coupled into L_1 (see Illustration 1.6). In this case, however, as $L_2 = 1$ H and $C_2 = 1$ F, the circuit $L_2 C_2$ is resonant at $\omega = 1$ rad/s, and therefore $Z_{22} = R_2 = 1 \, \Omega$ and only a pure resistance

$$r = \frac{\omega^2 k^2 L_1 L_2}{R_2} = 4k^2$$

is coupled into L_1.

The impedance Z_{1-2} is to be a pure resistance of 10 Ω. But as the circuit is parallel tuned, it is more convenient to state its admittance, and arrange for this to become a pure conductance of $\frac{1}{10}$ S. By inspection,

$$Y_{1-2} = \frac{r}{r^2 + \omega^2 L_1^2} - j \frac{\omega L_1}{r^2 + \omega^2 L_1^2} + j\omega C$$

$$= G_{1-2} + j B_{1-2}$$

The solution is now easily effected in two stages.

(a) Let it be assumed that C_1 is adjusted so that the susceptance of L_1 is annulled, and $B_{1-2} = 0$. Then $Y_{1-2} = G_{1-2}$, and for a pure resistance of 10 Ω to be presented at terminals 1–2, $Y_{1-2} = G_{1-2} = \frac{1}{10}$. Thus,

$$G_{1-2} = \frac{r}{r^2 + \omega^2 L_1^2} = \frac{r}{r^2 + 16} = \frac{1}{10}$$

or

$$r^2 - 10r + 16 = 0$$

whence

$$r = 8\,\Omega \quad \text{or} \quad 2\,\Omega$$

But $r = 4k^2$; and as k cannot exceed unity, only the solution $r = 2\,\Omega$ is valid, giving

$$k = \sqrt{\frac{r}{4}} = \frac{1}{\sqrt{2}}$$

(b) The susceptance at terminals 1–2 is annulled when

$$C_2 = \frac{L_2}{r^2 + \omega^2 L_2^2} = \frac{4}{4 + 16} = \frac{1}{5}\ \text{F}$$

The required values for the scaled transformer are therefore

$$k = \frac{1}{\sqrt{2}}, \quad C_1 = \frac{1}{5}\ \text{F}$$

Let α and β denote the impedance and frequency scaling factors, respectively. Then for the practical transformer operating between resistances of 10 kΩ and 1 kΩ at 1 MHz, $\alpha = 10^3$ and $\beta = 6 \cdot 28 \times 10^6$. The transformer parameters are therefore

$$L_2 = \frac{L_{2(s)}\alpha}{\beta} = \frac{1 \times 10^3}{6 \cdot 28 \times 10^6} \times 10^6 = 159 \cdot 2\ \mu\text{H}$$

$$L_1 = 4L_2 = 636 \cdot 8\ \mu\text{H}$$

$$C_2 = \frac{C_{2(s)}}{\alpha\beta} = \frac{1}{10^3 \times 6 \cdot 28 \times 10^6} \times 10^{12} = 159 \cdot 2\ \text{pF}$$

$$C_1 = \tfrac{1}{5}C_2 = 31 \cdot 84\ \text{pF}$$
$$k = 1/\sqrt{2}$$

Comment

(1) Observe the simplicity of the arithmetic in terms of the scaled values, and how easily the practical values are obtained from them.
(2) Note that the coupling coefficient is unaffected by scaling, since it is dimensionless (or alternatively, because the scale factors cancel out in the expression $k^2 = M^2/L_1L_2$).
(3) Note the adaptability of admittance to a problem of this kind.

ILLUSTRATION 3.8

Figure 3.6 incorporates a simplified small-signal representation for an amplifying device such as a valve or field-effect transistor. The transfer conductance of the device is $g_m = 5$ mA/V in the forward direction, but is assumed to be negligible in the reverse direction.

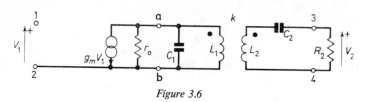

Figure 3.6

The parallel-series tuned transformer is to be designed to make $|V_2/V_1|$ a maximum at 1·591 MHz. $L_1 = 300$ μH, $L_2 = 100$ μH, L_2C_2 is in series-resonance at 1·591 MHz, and $R_2 = 1$ kΩ.

Determine the coupling coefficient k, the capacitors C_1 and C_2, and the voltage gain $|A| = |V_2/V_1|$ when r_0 is (a), 10 kΩ; (b) 100 kΩ.

Interpretation

For $|V_1|$ constant, $|A| = |V_2/V_1|$ is a maximum when $P_2 = |V_2|^2/R_2$ is a maximum. Assuming the transformer to have negligible dissipation, $P_2 = P_1$ where P_1 is the power entering it; and P_1 is a maximum when Z_{a-b} is a pure resistance equal to r_0. The problem is thus one of resistance matching, and the approach is similar to Illustration 3.7.

The calculations are simplified by scaling the circuit to $R_2 = 1$ Ω and $\omega = 1$ rad/s. For $f = 1·591$ MHz, $\omega = 10^7$ and $\beta = \omega = 10^7$; and for $R_2 = 1$ kΩ, $\alpha = R_2 = 10^3$. The transformer parameters are then

$$L_2 = 100 \times 10^{-6} \times \beta/\alpha = 1 \text{ H}$$

$$L_1 = 3L_2 = 3 \text{ H}$$

$$C_2 = 1/\omega^2 L_2 = 1 \text{ F}$$

while $R_2 = 1$ Ω, and $r_0 = 10$ Ω in case (a) and 100 Ω in case (b).

The scaled resistance coupled into L_1 is

$$r = \omega^2 k^2 L_1 L_2/R_2 = 3k^2$$

Following Illustration 3.7, k is found from the admissible (i.e. consistent with $k < 1$) value of r that satisfies the conductance equation

$$\frac{r}{r^2 + \omega^2 L_1^2} = \frac{1}{r_0} \tag{1}$$

and then C_1 is given by the susceptance-annullment equation

$$C_1 = \frac{L_1}{r^2 + \omega^2 L_1^2} \qquad (2)$$

which can be put also in the form

$$C_1 = L_1 / r_0 r$$

(a) When $r_0 = 10\,\Omega$, the conductance equation gives

$$r^2 - 10r + 9 = 0$$

whence

$$r = 9 \quad \text{or} \quad 1$$

The admissible root is $r = 1$, giving $k = 1/\sqrt{3}$ and $C_1 = 0{\cdot}3$ F, scaled, or 30 pF, actual.

(b) When $r_0 = 100\,\Omega$,

$$r^2 - 100r + 9 = 0$$

Since $(100)^2 \gg 4 \times 9$, the roots are extreme in values, the admissible one being very small.

Consider the general form

$$ar^2 + br + c = 0$$

for which

$$r = -\frac{b}{2a} \pm \frac{(b^2 - 4ac)^{1/2}}{2a}$$

When $b^2 \gg 4ac$, the binomial approximation gives

$$(b^2 - 4ac)^{1/2} \simeq b\left(1 - \frac{2ac}{b^2}\right)$$

Thus, as a close approximation,

$$r = -\frac{c}{b} \quad \text{or} \quad \frac{c}{b} - \frac{b}{a}$$

and in the present case,

$$r = 0{\cdot}09 \quad \text{or} \quad 99{\cdot}91$$

the value $r = 0{\cdot}09$ is the valid one, giving

$$k = \sqrt{0{\cdot}03} \quad \text{and} \quad C_1 = \frac{L_1}{r_0 r} = \frac{1}{3} \text{ F}, \quad \text{scaled, or 33·3 pF, actual.}$$

The voltage-gain calculation

Since the resistance looking-into terminals a–b of the transformer has been made equal to r_0 in each case, the net resistance between these terminals, including that of the shunt source, is $r_0/2$. The voltage between the terminals is therefore $g_m V_1 r_0/2$.

The powers entering and leaving the transformer (ignoring dissipation) are equal; and therefore,

$$\left[\frac{g_m V_1 r_0}{2}\right]^2 \cdot \frac{1}{r_0} = \frac{|V_2|^2}{R_2}$$

whence

$$|A| = \left|\frac{V_2}{V_1}\right| = \frac{g_m}{2}\sqrt{(r_0 R_2)}$$

The actual value of g_m is 5×10^{-3} S. Its scaled value is αg_m, or 5 S. The gains are therefore, in terms of scaled values,

(a), $$|A| = \tfrac{5}{2}\sqrt{(10\times1)} = 2\cdot5\sqrt{10}$$

and

(b), $$|A| = \tfrac{5}{2}\sqrt{(100\times1)} = 25$$

The same results are obtained in terms of the actual values: the scaled and actual gains are identical, so long as g_m is scaled as well as the circuit elements.

Comment

(1) Consider the expressions

$$\frac{1}{r_0} = \frac{r}{r^2+\omega^2 L_1^2}$$

and

$$C_1 = \frac{L_1}{r^2+\omega^2 L_1^2} = \frac{L_1}{r_0 r}$$

In case (b), for $r_0 = 100\ \Omega$ (scaled value), r^2 is only 1% of $\omega^2 L_1^2$ when $\omega = 1$. Then the expressions are closely approximated by

$$r_0 = \frac{\omega^2 L_1^2}{r} = \frac{\omega^2 L_1^2 R_2}{\omega^2 k^2 L_1 L_2} = \frac{L_1 R_2}{k^2 L_2}$$

and

$$C_1 = \frac{1}{\omega^2 L_1}$$

The first approximation gives the neat explicit expression for coupling coefficient,

$$k = \sqrt{\left[\frac{L_1 R_2}{r_0 L_2}\right]}$$

In case (a), however, for $r_0 = 10\,\Omega$, r^2 is 11% of $\omega^2 L_1^2$ and cannot be ignored in the denominators of equations (1) and (2). Its significance is demonstrated by comparing the values of C_1 for the two cases: for $r_0 = 100\,\Omega$, $C_1 = \frac{1}{3}\,\text{F}$; but for $r_0 = 10\,\Omega$, $C_1 = 0.9(\frac{1}{3})\,\text{F}$. The difference of 11% represents the error that would have been incurred if the approximation $C_1 = 1/\omega^2 L_1$ had been used for the case $r_0 = 10\,\Omega$: yet it is quite valid for $r_0 = 100\,\Omega$.

ILLUSTRATION 3.9

State the Thevenin–Helmholtz theorem and its dual, and explain, with simple illustrations, the considerations governing the choice of theorem for the solution of a particular network.

Interpretation

According to the Thevenin–Helmholtz theorem, any complicated network of linear circuit elements in association with sources of emf (or current) is equivalent, with respect to its output terminals, to a simple generator whose series impedance is equal to the output impedance of the network, and whose emf is equal to the potential difference between the output terminals when unloaded.

Some qualification of this statement is necessary in respect of contemporary devices. In the case of a network of intrinsically passive and linear LCR elements, the output impedance (or admittance) is calculable or measurable when all constant emf and current sources are set to zero, with their paths closed and open, respectively. But the theorem is also applicable to networks of a non-reciprocal kind, such as transistors and electron tubes, provided linearity may be assumed (as an approximation), and provided it is recognised that equivalent circuit representations for such devices must include current or voltage controlled constant emf or constant current sources (such as αI_b or $g_m V_{b'e}$ (hybrid π) in the case of a transistor, or μV_{gk} or $g_m V_{gk}$ in the case of a valve). See Illustrations 3.13, 3.14 and 3.16.

The dual of the Thevenin–Helmholtz theorem has been known as Norton's theorem, and states that, with respect to its output terminals, any complicated network of linear circuit elements in association with sources of current (or emf) is equivalent to a source of constant current acting in parallel with an admittance, the constant current being equal to that traversing the output terminals when short-circuited, and the admittance being equal to the output admittance of the network when rendered passive.

The choice of theorem is governed mainly by (1), the ease with which the equivalent generator can be found; and (2), the suitability of the type of generator for the topology of the network that follows it. But this point is less important than the first, for either form of equivalent generator is easily transformed into the other to afford optimum analysis (loop-current or nodal-voltage) of the network it feeds.

The practical value of both theorems depends on the ease with which the simple generators can be derived. This is governed by the topology of the network, the kinds of sources in it, and whether the branches are given as impedances or admittances. It is the current generator that is likely to be easier to derive if the network contains current sources and its topology is favourable for nodal-voltage analysis, and the emf generator if it contains emf sources and is favourable for loop-current (or mesh) analysis. These cases are illus-

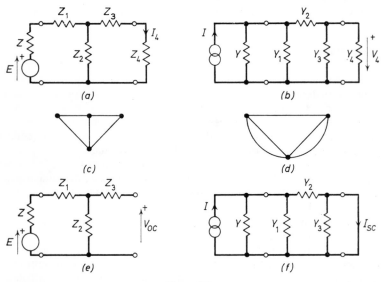

Figure 3.7

trated in Figure 3.7, in which (a) and (b) are T and π networks with sources and loads, (c) and (d) are the corresponding basic graphs (diagrams of basic conducting paths and junction points), while (e) and (f) show the networks under the open and and short-circuit conditions appropriate to finding the constant emf and constant-current forms of equivalent generator.

Examining the solution requirements formally, the graph for the T-network (closed-path source) contains 4 nodes and 5 branches. It therefore requires

$$L = B - N + 1 = 2 \quad \text{loop or mesh-current equations}$$

or

$$N - 1 = 3 \quad \text{nodal-voltage equations.}$$

Similarly, the π-network graph (open-path source) shows that 2 nodal voltage or 3 loop or mesh-current equations are required.

Mesh analysis is thus the optimum for T-network and nodal-voltage analysis for the π network. The networks shown are in fact duals, and the solutions are easily shown to be of the dual forms

$$I_4 = EZ_2/[(Z_3 + Z_4)(Z + Z_1 + Z_2) + Z_2(Z + Z_1)]$$

and

$$V_4 = IY_2/[(Y_3 + Y_4)(Y + Y_1 + Y_2) + Y_2(Y + Y_1)]$$

The optimum forms of analysis imply that the emf type equivalent generator is easier to derive by inspection for the T-network, and the current-type for the π. Referring to Figure 3.7(e), Z_3 is redundant in calculating the open-circuit voltage, and by inspection the emf of the equivalent generator is

$$E_0 = V_{0c} = \frac{EZ_2}{Z + Z_1 + Z_2}$$

while, with E reduced to zero (but its path left closed), the series impedance of the generator is given by series-parallel calculation as

$$Z_0 = Z_3 + \frac{Z_2(Z + Z_1)}{Z + Z_1 + Z_2}$$

Since Figure 3.7(f) is the dual, the parameters of the corresponding current generator could be written down merely by exchanging symbols. However, noting that Y_3 is redundant across a short-circuit, and that current divides between parallel admittances as does voltage across series impedances, the equivalent generator current is

$$I_0 = I_{sc} = \frac{IY_2}{Y + Y_1 + Y_2}$$

while, with the short-circuit removed and I reduced to zero but with its path left open, and noting that admittances in series combine like impedances in parallel (product/sum), the generator shunt admittance is

$$Y_0 = Y_3 + \frac{Y_2(Y + Y_1)}{Y + Y_1 + Y_2}$$

Comment

This illustration demonstrates the importance of examining the structure of a network before choosing a theorem for use as simplifying artifice: the value of an equivalence-theorem is lost or diminished unless the equivalence relationships are fairly obvious. In practice a π network is likely to be driven from an emf generator; but it is trivial to transform this into its constant-current equivalent.

The choice of theorem is, however, not governed solely by the topology of the circuit: one must consider, for example, the work involved in converting impedances into admittances or vice-versa, when these are complex quantities. The form of the data is thus an important additional consideration.

ILLUSTRATION 3.10

The measured open-circuit voltage at the output terminals of an active network of linear elements is $10\underline{/0}$ V. A pure resistance $r = 100\,\Omega$ is connected across the output terminals, and the voltage across it, measured in magnitude and phase relative to the first measurement, is 4·46 V, lagging by $\theta = \tan^{-1} 0\cdot5$ on the open-circuit voltage of the network. Determine the parameters of a simple generator equivalent to the network.

Interpretation

From the first measurement, the emf of the Thevenin-type equivalent generator is

$$E = V_{0c} = 10 \text{ V}$$

Let $Z = R+jX$ denote the equivalent generator impedance. Then from the second measurement,

$$V = \frac{Er}{r+Z} = \frac{10\times100}{100+R+jX} = 4\cdot46\underline{/-\theta}$$

where $\theta = \tan^{-1} 0\cdot5 = 26°34'$.

Inverting and transforming into real and imaginary components,

$$100+R+jX = 224 (\cos \theta+j \sin \theta)$$
$$= 224(0\cdot8944+j0\cdot4472)$$
$$= 200+j100$$

whence

$$R = 100\,\Omega \text{ and } X = 100\,\Omega$$

The equivalent Thevenin-type generator thus comprises an emf of magnitude 10 V in series with an impedance of $100+j100 \; \Omega$.

Comment

(1) If the frequency is known, the inductive element of the equivalent generator is $L = X/\omega = 100/\omega$, H.

(2) The data given is directly adaptable to formulate the Thevenin-type generator. The constant-current equivalent is, however, easily obtained from this. It comprises a constant current

$$I = \frac{E}{Z} = \frac{1}{10+j10} = \frac{1}{10\sqrt{2}} \underline{/-45°} \; A$$

in parallel with an admittance

$$Y = \frac{1}{Z} = G - jB = \frac{1}{200} - j\frac{1}{200} \; S$$

Note the shift in phase of this current relative to the open-circuit voltage of the network.

ILLUSTRATION 3.11

State the Helmholtz–Thevenin theorem for ac networks involving a number of sources and interconnections.

A network of impedances and sources of alternating emf's has two output terminals. The open-circuit voltage at the terminals is 260 V. The current flowing when the terminals are short-circuited is 20 A, and 13 A when connected through a coil of 11 Ω reactance and negligible resistance.

Determine the components of the equivalent circuit feeding the terminals. What value of load impedance would give maximum power output?

(L.U., Part 2, Electrical Theory and Measurements).

Interpretation

(a) Statement: see Illustration 3.9.

(b) The problem is similar in principle to Illustration 3.10, but only magnitudes are involved.

The emf of the equivalent Thevenin generator (equivalent circuit) is $E = V_{0c} = 260$ V. Let $|Z| = \sqrt{(R^2+X^2)}$ denote the magnitude of its impedance. Then for the two conditions given,

$$\frac{E}{\sqrt{(R^2+X^2)}} = \frac{260}{\sqrt{(R^2+X^2)}} = 20 \text{ A}$$

whence $R^2 = 169 - X^2$, and

$$\frac{E}{\sqrt{\{R^2+(X+11)^2\}}} = \frac{260}{\sqrt{\{169-X^2+(X+11)^2\}}} = 13$$

whence

$$X = 5\,\Omega, \quad R = 12\,\Omega \quad \text{or} \quad Z = 12+j5\,\Omega$$

(c) The maximum power is absorbed from the network or its equivalent Thevenin generator when the load impedance is the conjugate of the generator impedance, i.e. $12-j5$ (see Illustration 4.2).

Comment

It is interesting to note that the correct sign of the reactance is deducible from magnitudes only.

Note also that the theorem is not restricted to sinusoidal ac sources.

ILLUSTRATION 3.12

Discuss briefly the meaning of the terms 'constant voltage source' and 'constant current source', and show that a practical voltage source can be replaced by an equivalent current source.

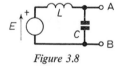

Figure 3.8

Figure 3.8 shows a sinusoidal voltage source. Draw a circuit diagram showing the equivalent current source between terminals A and B. Find the strength of the source and the value of its shunt admittance in terms of ω, E, L and C.

If $E = 1$V and $\omega = 10^6$ rad/s find the values of L and C to make the circuit of Figure 3.8 equivalent to an unshunted constant-current source of 1 mA.

(L. U. Part 2, Electrical Theory and Measurements).

Interpretation

The equivalence between current and voltage sources is fully illustrated in Chapter 1.

Referring to Figure 3.8, the equivalent current source comprises a constant current I_{AB} equal to that traversing terminals AB under a short-circuit condition, in parallel with an admittance Y_{AB} equal to that between terminals AB when E is reduced to zero but with its path closed. Thus,

$$I_{AB} = E/j\omega L, \quad Y_{AB} = j(\omega C - 1/\omega L)$$

The constant current has a magnitude of 1 mA when

$$\frac{E}{\omega L} = \frac{1}{10^6 L} = 10^{-3} \text{ A}$$

whence

$$L = 1 \text{ mH}$$

It behaves as an unshunted constant current when

$$Y_{AB} = j(\omega C - 1/\omega L) = 0$$

whence

$$C = 1/\omega^2 L = 1000 \text{ pF}$$

Comment

This simple but instructive problem brings out the theoretical possibility of realising an ideal infinite impedance (zero admittance) current source by utilising the effects of resonance between ideal elements having infinite Q-factors. In practice, however, the inductor L would be dissipative, while the emf source E would also have a significant series resistance (unless very closely regulated). The reader might re-examine the problem when L has a finite Q-factor, and the emf source has a series resistance r: the shunt admittance could be low (r small, Q high), but never zero.

ILLUSTRATION 3.13

By means of the Thevenin–Helmholtz theorem, obtain an equivalent emf generator to represent a triode valve under small-signal conditions in (a), the common-anode (cathode-follower) circuit; and (b), the common-grid (grounded-grid) circuit.

Interpretation

It is assumed that the grid current in each case is zero, and that the operating conditions are linear.

(a) The common-anode circuit is represented in basic form in Figure 3.9(a). Figure 3.9(b) is an equivalent circuit for calculating the open circuit voltage; and Figure 3.9(c) is for calculating the output resistance when the input V_1 is reduced to zero.

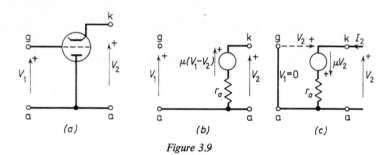

Figure 3.9

From Figure 3.9(b) the emf of the equivalent generator is the open-circuit value of V_2, or

$$E_0 = V_2 = \mu(V_1 - V_2) = \frac{\mu}{1+\mu} V_1$$

The resistance of the generator is the output resistance of the circuit when $V_1 = 0$. This is found by applying V_2 as a source and finding the current I_2 it produces. Thus, from Figure 3.9(c),

$$I_2 = V_2(1+\mu)/r_a$$

whence

$$R_0 = \frac{V_2}{I_2} = \frac{r_a}{1+\mu}$$

The common-anode circuit is thus equivalently represented by a simple generator of emf $E_0 = \mu V_1/(1+\mu)$ and series resistance $R_0 = r_a/(1+\mu)$.

(b) The common-grid circuit is shown in Figure 3.10(a). In this case the output impedance is a function of the impedance of the input source, which must therefore be included in the calculations.

From Figure 3.10(b), noting that no current is drawn from the source E_1, Z_1 under the open-circuit condition, the equivalent generator emf is

$$E_0 = V_2 = (1+\mu)V_1 = (1+\mu)E_1$$

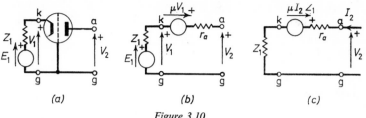

Figure 3.10

From Figure 3.10(c), in which V_2 is applied as a source while E_1 is reduced to zero,

$$V_2 = I_2 r_a + \mu I_2 Z_1 + I_2 Z_1$$

whence the equivalent generator series impedance is

$$Z_0 = \frac{V_2}{I_2} = r_a + Z_1(1+\mu)$$

Comment

These examples emphasise the importance of recognising that the impedance between two points in a network is fundamentally the relationship between a voltage between the points and the current it produces.

While in the case of a passive reciprocal network the impedance or admittance between two points can often be calculated directly by compounding branches (as in Illustration 3.9, for example), a direct Ohm's law approach is essential in the case of non-reciprocal (often called *active*) devices such as valves and transistors; for in such cases, activation from an external source of voltage or current causes a proportionate source to appear within the network. Such a *controlled source* is exemplified by $\mu I_2 Z_1$ in Figure 3.10(c); and the effective impedance to the impressed voltage V_2 is quite different from that which might be inferred erroneously from an inspection of the circuit under passive (unactivated) conditions.

ILLUSTRATION 3.14

A non-reciprocal two-port device (which could be a transistor), assumed linear and operated under small-signal conditions, is defined by either of the matrices,

$$[Z] = \begin{bmatrix} Z_{11} & Z_{12} \\ Z_{21} & Z_{22} \end{bmatrix} \quad \text{or} \quad [h] = \begin{bmatrix} h_{11} & h_{12} \\ h_{21} & h_{22} \end{bmatrix}$$

118

A generator of emf E_1 and series impedance Z_1 is connected to the input terminals. Find whichever of the simple equivalent generators is the easier to derive for each matrix, as a substitute for the activated device with respect to its output terminals.

Interpretation

The basic arrangement is indicated in Figure 3.11, with polarity orientations in conformity with the standard two-port convention.

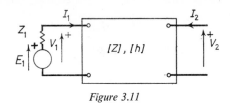

Figure 3.11

(a) For $[Z]$,

$$\begin{bmatrix} Z_{11} & Z_{12} \\ Z_{21} & Z_{22} \end{bmatrix} \cdot \begin{bmatrix} I_1 \\ I_2 \end{bmatrix} = \begin{bmatrix} V_1 \\ V_2 \end{bmatrix} = \begin{bmatrix} E_1 - I_1 Z_1 \\ V_2 \end{bmatrix}$$

or,

$$(Z_{11}+Z_1)I_1+Z_{12}I_2 = E_1$$
$$Z_{21}I_1+Z_{22}I_2 = V_2$$

The constant-current (Norton) equivalent generator requires evaluation of I_2 when $V_2 = 0$ (short-circuit condition), and solution for both I_1 and I_2 would be required. On the other hand, putting $I_2 = 0$ gives V_2 as the open-circuit voltage, and this is immediately the emf E_0 of the equivalent Thevenin-type generator with only I_1 to be eliminated. When $I_2 = 0$, $I_1 = E_1/(Z_1+Z_{11})$ and

$$E_0 = V_2|_{I_2 = 0} = E_1 \cdot \frac{Z_{21}}{Z_1+Z_{11}}$$

The equivalent-generator impedance Z_0 is found by putting $E_1 = 0$ and treating V_2 as an applied voltage. Then, eliminating I_1,

$$-\frac{Z_{21}Z_{12}}{Z_1+Z_{11}} \cdot I_2+Z_{22}I_2 = V_2$$

whence

$$Z_0 = \frac{V_2}{I_2}\bigg|_{E_1 = 0} = Z_{22}-\frac{Z_{21}Z_{12}}{Z_1+Z_{11}} = Z_{22}-\frac{Z_{12}E_0}{E_1}$$

(b) [h] leads to

$$(h_{11}+Z_1)I_1+h_{12}V_2 = E_1$$
$$h_{21}I_1+h_{22}V_2 = I_2$$

In the case putting $V_2 = 0$ gives I_2 as the short-circuit current, which is immediately the current I_0 in the constant-current (Norton) generator representation. Thus by simple manipulation,

$$I_0 = I_2\big|_{V_2=0} = E_1 \cdot \frac{h_{21}}{h_{11}+Z_1}$$

and, putting $E_1 = 0$ and treating I_2 as a current source,

$$Y_0 = \frac{I_2}{V_2}\bigg|_{E_1=0} = h_{22} - \frac{h_{21}h_{12}}{h_{11}+Z_1} = h_{22} - \frac{h_{12}I_0}{E_1}$$

Comment

Illustration 3.9 shows that the perspicuity of an equivalent generator is influenced by the topology of the active network. In this example the topology is not known, but the external behaviour is specified instead. The perspicuous equivalent generator is then governed by the type of two-port equation; for on this rests the choice between finding an open-circuit voltage or a short-circuit current.

For transistors it is customary to use letter subscripts for h-matrix elements. These identify the senses of the parameters, and the configuration to which they relate. For example, the elements for the common-emitter circuit are symbolised by h_{ie}, h_{re}, h_{fe}, h_{oe}, in place of h_{11}, h_{12}, h_{21}, h_{22}.

As an example of application, consider a transistor having common-base h-parameters $h_{ib} = 21 \cdot 6\,\Omega$, $h_{rb} = 2 \cdot 9 \times 10^{-4}$, $h_{fb} = -0 \cdot 98$, $h_{ob} = 0 \cdot 49 \times 10^{-6}$ S. For an input voltage V_1 between emitter and base,

$$I_0 = V_1 h_{fb}/h_{ib} = -4 \cdot 54 \times 10^{-2} V_1$$
$$Y_0 = h_{ob} - h_{rb}I_0/V_1 = 13 \cdot 7 \times 10^{-6} \text{ S}$$

Then for a collector-base load of resistance $R_2 = 2$ kΩ, taking V_1 and V_2 as voltage rises as for the two-port convention,

$$\frac{V_2}{V_1} = \frac{-I_0}{Y_0+G_2} = \frac{4 \cdot 54 \times 10^{-2} V_1}{5 \cdot 14 \times 10^{-4} V_1} = 88 \cdot 4.$$

It is not very easy to derive either simple equivalent generator by reduction of a transistor circuit-model (such as the T or hybrid π);

but it is easy to derive both from data in matrix form, particularly when the elements may be regarded as wholly real, as at low frequencies; but with no special difficulty even when they are complex.

The replacement of a transistor by an equivalent generator is a valuable simplification, subject to reasonable constancy of parameters, for studies in which the load, which may be complex, is the variable; and also for an iterative approach to multi-stage amplifier gain calculations. The substitution of an equivalent generator does, however, conceal the input impedance of the device; and this may be important in relation to the source.

ILLUSTRATION 3.15

A transistor, assumed linear and operated under small-signal conditions as a two-port device, is defined by an h-matrix according to the customary two-port polarity conventions. A generator of emf E_1 and series impedance Z_1 is connected to the input port. Find expressions for the elements I_0 and Y_0 of a simple current generator that may replace the activated transistor at its output port.

A two stage common-emitter transistor amplifier comprises two identical stages, each with a collector load resistance $R_c = 5$ kΩ. The h-parameters for each transistor are $h_{ie} = 1100\,\Omega$, $h_{re} = 2\cdot5\times10^{-4}$ $h_{fe} = 50$, $h_{oe} = 25\times10^{-6}$ S. Calculate, without approximations, the voltage gain V_2/V_1, where V_1 is the input voltage between base and emitter of the first stage and V_2 is the output voltage of the second stage between collector and emitter.

(P.C.L., BSc. (Hons.), C.N.A.A., Electronics, Year 2)

Interpretation

(a) See Illustration 3.14.

(b) The first stage is replaceable by $I_{01} = V_1 h_{fe}/h_{ie} = 4\cdot54\times10^{-2}V_1$ and $Y_{01} = h_{oe}-(h_{re}I_{01}/V_1)+1/R_c = 2\cdot14\times10^{-4}$ S. This current generator transforms into an emf generator $E_{01} = -I_{01}/Y_{01} = -212V_1$ (orientation for rise) and $Z_{01} = 1/Y_{01} = 4680\,\Omega$. This is the equivalent generator driving the second stage.

For the second stage, including the collector load resistor R_c,

$$I_{02} = E_{01}h_{fe}/(Z_{01}+h_{ie}) = -1\cdot83V_1$$

$$Y_{02} = h_{oe}-(h_{re}I_{02}/E_{01})+1/R_c = 2\cdot23\times10^{-4}\ \text{S}$$

Then V_2, oriented for rise, is $V_2 = -I_{02}/Y_{02} = 1\cdot83V_1/2\cdot23\times10^{-4}$ and the voltage gain is therefore

$$V_2/V_1 = 1\cdot83\times10^4/2\cdot23 = 8200$$

Comment

The calculations are routine, very simple yet free from approxima-tions, and extensible to more stages. Other exact methods, such as chain-matrix multiplication, would be much more tedious.

ILLUSTRATION 3.16

In Figure 1.10, Illustration 1.5, $Z_g = 1\,\text{k}\Omega$, pure resistance; $R_1 = 15\,\text{k}\Omega$; $R_2 = 10\,\text{k}\Omega$; $R_c = 3\,\text{k}\Omega$; $Z_L = 3\,\text{k}\Omega$, pure resistance; and the transistor has the parameters listed.

(a) Simplify the equivalent circuit of Figure 1.11(a) and calculate the voltage gains V_0/V_i for a frequency of order 1 kHz, assuming that the reactances of C_1, C_2 and C_3 are negligible.

(b) Estimate the upper 3 dB frequency f_2 by resolving the capaci-tances into an approximately equivalent single one between b' and e (as in Fig. 2.5, Illustration 2.7).

Interpretation

(a) $g_{b'c} \ll g_m$ and the omission of $g_{b'c}$ is therefore justified; and at a frequency of the order stated the capacitances have negligible effect and may also be excluded. In addition to the transistor parameters, $Y_g = 1/Z_g = 10^{-3}\,\text{S}$; $G = 1/R_1 + 1/R_2 = 1\cdot67 \times 10^{-4}\,\text{S}$; $G_c = 1/R_c = 3\cdot33 \times 10^{-4}\,\text{S}$; and $Y_L = 1/Z_L = 3\cdot33 \times 10^{-4}\,\text{S}$.

For the calculation of V_0/V_i, where V_i is the input terminal voltage, I_g, Y_g and G are redundant. Then,

$$V_{b'e} = V_i r_{b'e}/(r_{bb'} + r_{b'e}) = 0\cdot957 V_i$$
$$V_0 = -g_m V_{b'e}/(g_{ce} + G_c + Y_L) = -55\cdot6 V_{b'e}$$

and

$$V_0/V_i = -56\cdot6 \times 0\cdot957 = -54\cdot1$$

For the calculation of V_0/E_g, the circuit elements up to and including $g_{b'e}$ may be replaced by a simple equivalent current generator, compris-ing a current I_{eq} acting from e to b' in parallel with a conductance G_{eq}. Noting that $I_g = E_g/Z_g = 10^{-3}E_g$ and referring to Illustration 3.9, the generator components are

$$I_{eq} = \frac{I_g g_{bb'}}{Y_g + G + g_{bb'}} = \frac{10^{-3}E_g \times 9\cdot1 \times 10^{-3}}{10\cdot27 \times 10^{-3}} = 8\cdot86 \times 10^{-4}E_g$$

$$G_{eq} = g_{b'e} + \frac{g_{bb'}(Y_g + G)}{g_{bb'} + Y_g + G} = 1\cdot43 \times 10^{-3}\,\text{S}$$

Then $V_{b'e} = I_{eq}/G_{eq} = 0.62E_g$, and

$$V_0/E_g = -g_m V_{b'e}/(g_{ce}+G_c+Y_L)E_g = -35.1$$

(b) At the 3 dB frequency f_2 it is a fair approximation to assume that $V_0/V_{b'e}$ is not greatly affected by $C_{b'e}$, as $\omega_2 C_{b'c} \ll g_m$. Thus,

$$V_0/V_{b'e} \simeq -g_m(g_{ce}+G_c+Y_L) \simeq -56$$

Then at $b'e$,

$$C_{eq} = C_{b'e} + C_{b'c}(1+56) \simeq 1000 \text{ pF}$$

At f_2, $|V_{b'e}|$ is divided by $\sqrt{2}$, and this is so when

$$\omega_2 C_{eq} = G_{eq}, \quad \text{whence} \quad f_2 = 228 \text{ kHz}.$$

Comment

The low frequency gain $V_0/V_i = -54.1$ is remarkably close to -54.02, obtained without approximations by solving the nodal-voltage equations in Illustration 1.5 with a digital computer. This certainly validates the exclusion of $g_{b'c}$ from the gain calculations. But its omission is not necessarily justified for input admittance calculations, which are especially sensitive to a parallel feedback element (see Illustration 6.6). In this instance, taking $V_0/V_{b'e}$ as -57, the equivalent conductance at $b'e$ is

$$G_{eq} = g_{b'e} + g_{b'c}(1+57) = (4+0.29)10^{-4} = 4.29 \times 10^{-4} \text{ S}$$

so that $g_{b'c}$ has an influence of about 7%.

The frequency characteristic of the amplifier determined with the computer is shown in the following table.

f, kHz	0.1	1.0	10	100	1000	10 000
$|V_0/V_i|$	54.02	54.02	54.01	53.87	43.77	7.02

A plot of this frequency response shows that f_2 is roughly 1.5 MHz. Using the approximations illustrated in part (b), when $Z_g = 0$ so that $V_i = E_g$, $G_{eq} = g_{bb'} + g_{b'e} = 9.5 \times 10^{-3}$ and $f_2 \simeq 1.5$ MHz. This is therefore of the right order, and the approach is justified. It is important to observe the sensitivity of f_2 to source resistance, for when $Z_g = 1 \text{ k}\Omega$ f_2 is only about 230 kHz. In practice, it is often sufficient to know only the order of f_2.

ILLUSTRATION 3.17

(a) State and verify the reciprocity theorem.
(b) The external relations for a linear four-terminal network may be expressed in the form

$$V_1 = AV_2 - BI_2$$
$$I_1 = CV_2 - DI_2$$

where V_1, I_1 and V_2, I_2 denote the input and output voltages and currents, while A, B, C, and D are constants for the network. Show, by means of the reciprocity theorem, that if the network is passive and bilateral, $AD - BC = 1$.

Interpretation

(a) Let the current at a point y in a network be due to an emf at a point x. Then, provided the network comprises linear and bilateral elements only, the same current would flow at the point x if the emf were transferred to the point y.

This statement is consistent with the symmetry of the network determinant (see Illustration 1.2). The current in the mth mesh of a generalised linear network due to one emf in the kth mesh is

$$I_m = E_k \Delta_{km} / \Delta$$

Similarly, when the emf and current positions are interchanged,

$$I_k = E_m \Delta_{mk} / \Delta$$

But for a bilateral linear system, $\Delta_{km} = \Delta_{mk}$ and therefore $I_k = I_m$ when $E_m = E_k$.

(b) Since A, B, C, and D are constants at a given frequency for the network, they are invariable with external conditions and therefore valid for short-circuit conditions. Let the output terminals be short-circuited. Then $V_2 = 0$, and from the first equation,

$$I_2 = -V_1 / B$$

Now let the input terminals be short-circuited while V_2 is applied to the output terminals. Then $V_1 = 0$, and the equations give

$$I_2 = AV_2 / B$$
$$I_1 = CV_2 - DI_2 = V_2(C - AD/B)$$

As the network is linear, passive and bilateral, it conforms to the reciprocity theorem. Thus, if $V_2 = V_1$, where V_1 is the voltage that was applied when the output terminals were short-circuited,

$$I_1 = V_2(C - AD/B) = I_2 = -V_1/B = -V_2/B$$

whence

$$AD - BC = 1.$$

Comment

The reciprocity theorem is of great fundamental importance, for it has implications in all network calculations. Nevertheless, it is consciously invoked much more often as a powerful premise for theoretical reasoning than as an expedient for the practical solution of a network (as compared, for example, with the Thevenin or superposition theorems). It is important to note that the theorem is valid only for an interchange in the points of action of emf and current: it is not valid for an interchange that includes also the impedances existing at the points concerned, unless they happen to be equal.

Violation of the reciprocity theorem is characteristic of a network containing thermionic valves, rectifiers, transistors, etc.

ILLUSTRATION 3.18

Two linear, passive reciprocal networks have the same number of accessible terminals but different internal structures. State a general requirement for the networks to exhibit identical external behaviour at a given frequency.

Three terminals 1, 2, 3 may be connected respectively to admittances Y_1, Y_2, Y_3 meeting at a point in the form of a star, or be the vertices of a triangle of admittances Y_a, Y_b, Y_c linking respectively terminal 3 with 1, 1 with 2 and 2 with 3. Find relationships between the sets of admittances for the networks to be externally equivalent at a given frequency.

Interpretation

The external behaviour of two networks having the same number of accessible terminals but different internal structures are identical at a given frequency when their indefinite admittance (or impedance) matrices are equal.

In the case of the star and delta networks, which are basic arrangements for the interconnection of three accessible terminals, it is sufficient to form three independent equations by equating three corresponding elements in the 3×3 indefinite matrices. Referring to equation (18), Introduction, Chapter 1, the nodal admittances Y_{11}, Y_{22}, Y_{33} are the easiest ones to find. Let the generalised 3-terminal unit in Figure 1.4(b) be occupied by the star Y_1, Y_2, Y_3. Noting that admittances in series combine in the form $Y_iY_k/(Y_i+Y_k)$,

$$Y_{11_{star}} = \frac{I_1}{V_1}\bigg|_{V_2 = 0,\ V_3 = 0} = \frac{Y_1(Y_2+Y_3)}{\Sigma Y} = \frac{Y_1Y_2}{\Sigma Y} + \frac{Y_1Y_3}{\Sigma Y}$$

$$Y_{22_{star}} = \frac{I_2}{V_2}\bigg|_{V_1 = 0,\ V_3 = 0} = \frac{Y_2(Y_1+Y_3)}{\Sigma Y} = \frac{Y_1Y_2}{\Sigma Y} + \frac{Y_2Y_3}{\Sigma Y}$$

$$Y_{33_{star}} = \frac{I_3}{V_3}\bigg|_{V_1 = 0,\ V_2 = 0} = \frac{Y_3(Y_1+Y_2)}{\Sigma Y} = \frac{Y_1Y_3}{\Sigma Y} + \frac{Y_2Y_3}{\Sigma Y}$$

Replacing the star by the delta,

$$Y_{11_{delta}} = Y_a+Y_b, \quad Y_{22_{delta}} = Y_b+Y_c, \quad Y_{33_{delta}} = Y_a+Y_c$$

Observing that each term appears twice on the right hand side in each set of equations, equating like matrix elements gives the obvious solutions,

$$Y_a = Y_1Y_3/\Sigma Y, \quad Y_b = Y_1Y_2/\Sigma Y, \quad Y_c = Y_2Y_3/\Sigma Y \qquad (1)$$

Transformation from delta to star gives the dual of equation (1), in terms of impedance Z_1, Z_2, Z_3 and Z_a, Z_b, Z_c and open-circuit mesh-impedance elements Z_{11}, Z_{22}, Z_{33}. Equating impedances at each corresponding terminal-pair under open-circuit conditions at the others gives

$$Z_1 = Z_aZ_b/\Sigma Z, \quad Z_2 = Z_bZ_c/\Sigma Z, \quad Z_3 = Z_aZ_c/\Sigma Z \qquad (2)$$

Comment

(1) All other matrix elements are equal under the equivalence conditions. For example,

$$Y_{32_{delta}} = \frac{I_3}{V_2}\bigg|_{V_1 = 0,\ V_3 = 0} = -Y_c = -\frac{Y_2Y_3}{\Sigma Y} = Y_{32_{star}}$$

(Let V_0^+ be the potential of the star point. When nodes 1 and 3 are short-circuited to the datum, $V_0\Sigma Y - V_2Y_2 = 0$, $I_3 = -Y_3V_0 = -Y_3Y_2V_2/\Sigma Y$ and therefore $Y_{32} = I_3/V_2 = -Y_2Y_3/\Sigma Y$).

(2) The transformations reduce the topological complexity of a network. Star to delta eliminates a node (the star point), and delta to star eliminates a loop. But the advantage gained by reducing the nodal voltage or loop-current equations is off-set by the arithmetic of the

transformations, especially when the immittances are complex. Note that the equivalences are not always physically realisable.

(3) Equation (1) is a special case of a general theorem, sometimes called Rosen's theorem, according to which a star of n admittances is replaceable by a pair-connected system in which admittances of the form $Y_i Y_k / \Sigma Y$ link each terminal $(1, 2, \ldots, i, \ldots, k, \ldots, n)$ to every other one.

ILLUSTRATION 3.19

A four-terminal non-reciprocal network is represented in the form of a ladder structure having input terminals 1, 2 and output terminals 3, 4. 2 and 4 are common. The branches, from input side to output, are respectively a series admittance Y_1, a shunt admittance Y_2, a series admittance Y_3 and a shunt admittance Y_4 in parallel with a current-source $g_m V_1$. This flows away from terminal 3 for an input voltage V_1 oriented towards terminal 1. Find the Y-matrix.

Interpretation

A neat approach is to apply the star-delta transformation to the branches Y_1, Y_2 and Y_3, which reduces the topology to that of a π-network for which the Y-matrix is easily found by inspection (see, for example, Illustration 1.16). Applying equation (1) Illustration 3.17, gives

$$[Y] = \begin{bmatrix} \dfrac{Y_1(Y_2 + Y_3)}{\Sigma Y} & -\dfrac{Y_1 Y_3}{\Sigma Y} \\[3mm] g_m - \dfrac{Y_1 Y_3}{\Sigma Y} & Y_4 + \dfrac{Y_3(Y_1 + Y_2)}{\Sigma Y} \end{bmatrix}$$

where

$$\Sigma Y = Y_1 + Y_2 + Y_3$$

Comment

Note that g_m is readily placed in the above matrix by superposition, as explained in Illustration 116.

ILLUSTRATION 3.20

(a) Show that the arms of a lattice network are related in a simple way to the half-section of a symmetrical two-port network equivalent to it.

(b) A lattice network has $0{\cdot}5\ \mu\text{F}$ capacitors for its line or series arms C_x, $600\ \Omega$ resistors for its lattice or cross arms R_y, and its output terminals are closed through a $400\ \Omega$ resistor R_t. Consider three ways in which the input impedance might be found and evaluate it by the simplest method at $\omega = 5000\ \text{rad/s}$.

Interpretation

Two linear networks with specified input and output ports are equivalent in behaviour between these ports when their corresponding two-port matrices are equal. As a 2×2 reciprocal immittance matrix is realisable (though not necessarily physically) in the form of a T or π network, these may be regarded as basic equivalents to any passive, linear two-port network, regardless of its internal complexity. Figure 3.12(b) is a symmetrical and balanced form of the π-network representing any symmetrical and balanced two-port network to be set equivalent to the lattice network of Figure 3.12(a).

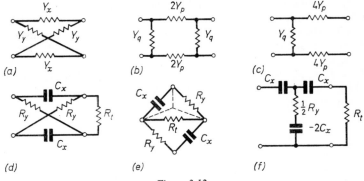

Figure 3.12

As the networks are linear, passive and symmetrical, $Z_{12} = Z_{21}$, $Z_{11} = Z_{22}$, $Y_{12} = Y_{21}$ and $Y_{11} = Y_{22}$. Two matrix elements are therefore sufficient to establish equivalence, and $[Y]$ is convenient and instructive.

Figure 3.12(a) decomposes into two paralleled networks; one comprising the series elements Y_x only, and the other the crossed elements Y_y. Then,

$$[Y] = \begin{bmatrix} \frac{1}{2}Y_x & -\frac{1}{2}Y_x \\ -\frac{1}{2}Y_x & \frac{1}{2}Y_x \end{bmatrix} + \begin{bmatrix} \frac{1}{2}Y_y & \frac{1}{2}Y_y \\ \frac{1}{2}Y_y & \frac{1}{2}Y_y \end{bmatrix}$$

$$= \begin{bmatrix} \frac{1}{2}(Y_y+Y_x) & \frac{1}{2}(Y_y-Y_x) \\ \frac{1}{2}(Y_y-Y_x) & \frac{1}{2}(Y_y+Y_x) \end{bmatrix}$$

For equivalence with Figure 3.12(b), for which Y_{11} and Y_{12} are obvious,

$$\tfrac{1}{2}(Y_y+Y_x) = Y_q+Y_p$$
$$\tfrac{1}{2}(Y_y-Y_x) = -Y_p$$

whence, adding and subtracting,

$$Y_y = Y_q, \quad Y_x = Y_q+2Y_p$$

But these are recognisable as the admittances looking into the half-section of Figure 3.12(c) under, respectively, open-circuit and short-circuit conditions. This verifies *Bartlett's bisection theorem*, according to which the line and cross arms of a lattice network equivalent to a given symmetrical two-port network are equal, respectively, to the input impedances or admittances of the bisected network under short-circuit and open-circuit conditions at the point of bisection.

(b) The input impedance may be found by mesh analysis, or by compounding branches after making the delta to star transformation indicated. Both methods are tedious. It is easiest to replace the lattice by a T-network (structural balance is inconsequential in this instance) of series branches Z_p and a shunt branch Z_q. The half-section has branches Z_p and $2Z_q$; and by Bartlett's theorem $Z_x = Z_p$ and $Z_y = Z_p+2Z_q$, or $Z_p = Z_x$ and $Z_q = (Z_y-Z_x)/2$. This gives Figure 3.12(f), in which $C_x = 1$ F and $R_y = \tfrac{3}{2}\,\Omega$ when scaled to $R_t = 1\,\Omega$ and $\omega = 1$ rad/s. Directly, writing $s = j\omega$,

$$Z_{\text{in}_{\text{scaled}}} = \frac{1}{s} + \frac{(1+1/s)(3-2/s)}{7+2/s}$$

Putting $s = j1$ gives the scaled value of $1.17\underline{/-53°}\,\Omega$, and the actual value for $\omega = 5000$ and $R_t = 400\,\Omega$ as $468\underline{/-53°}\,\Omega$.

Comment

(1) The negative capacitor in Figure 3.12(f) means that the structure is unrealisable physically in passive form. It is nevertheless mathematically valid.

(2) The limitations on the equivalence between Figure 3.12(a) and (b) are shown by comparing their indefinite admittance matrices, which can be written by inspection as there are no internal nodes (in such a case $Y_{kk} = \Sigma Y$ at a node k, and Y_{ij} is the admittance linking a pair of nodes i and j). It will be seen that they cannot be identical.

ILLUSTRATION 3.21

Obtain the indefinite impedance matrices for Figures 3.13(a) and (b) and thence show that two pure inductors coupled by pure mutual inductance and connected at one end may be represented equivalently by a star or delta of three self-inductors.

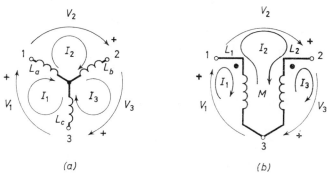

(a)　　　　　　　　(b)

Figure 3.13

Interpretation

The indefinite impedance matrix $[Z]_i$ for a star like Figure 3.13(a) (but not restricted to three branches) is the mesh-impedance matrix when all ports of the network are energised with voltages whose polarities are consistent with mesh currents of common orientation, as indicated. Its form is then dual with that of the indefinite admittance matrix for a pair-connected network (such as the delta in Illustration 3.18), and its elements are open-circuit impedances of general form $Z_{ik} = V_i/I_k$ when all currents except the kth. are set to zero. While the elements for Figure 3.13(a) are obvious, they are exemplified formally by

$$Z_{11} = \frac{V_1}{I_1}\bigg|_{I_2 = 0,\ I_3 = 0} = s(L_a + L_c); \quad Z_{32} = \frac{V_3}{I_3}\bigg|_{I_1 = 0,\ I_2 = 0} = -sL_b$$

Thus,

$$[Z]_{i(a)} = s \begin{bmatrix} L_a + L_c & -L_a & -L_c \\ -L_a & L_a + L_b & -L_b \\ -L_c & -L_b & L_b + L_c \end{bmatrix} \quad (1)$$

For Figure 3.13(b), noting that I_2 traverses both coils and that every current traversing one coil induces into the other a voltage whose polarity must be related to the self-impedance voltage falls according to the dot positions assumed (see Illustration 1.6),

$$sL_1I_1 - s(L_1 - M)I_2 - sMI_3 = V_1$$
$$-s(L_1 - M)I_1 + s(L_1 + L_2 - 2M)I_2 - s(L_2 - M)I_3 = V_2$$
$$-sMI_1 - s(L_2 - M)I_2 + sL_2I_3 = V_3$$

and therefore

$$[Z]_{i(b)} = s \begin{bmatrix} L_1 & -L_1 + M & -M \\ -L_1 + M & L_1 + L_2 - 2M & -L_2 + M \\ -M & -L_2 + M & L_2 \end{bmatrix} \qquad (2)$$

Comparing eqn. (1) with (2) shows that the matrices are equal when

$$L_a = L_1 - M, \quad L_b = L_2 - M, \quad L_c = M \qquad (3)$$

The equivalence is thus physically realisable for the winding-senses assumed, provided $M < (L_1, L_2)$. The star L_a, L_b, L_c, is readily transformed into an equivalent delta by means of equation (1), Illustration 3.17.

Comment

The replacement of three physical inductors by two with mutual inductance is an economy. Sometimes it may validate physical realisation. This is exemplified by Figure 3.14(a), in which the negative

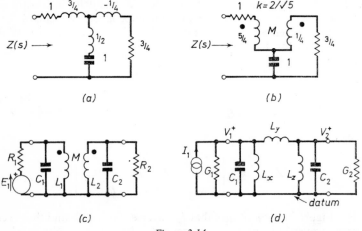

Figure 3.14

inductor of $-\frac{1}{4}$ H precludes physical realisation in passive form until the three inductors are replaced by two positive ones with mutual inductance, as in Figure 3.14(b).

The transformations are useful in analysis. The formulation of nodal-voltage equations is awkward for circuits containing mutual inductors, but is routine when these are replaced by deltas of self inductors which need not be physically realisable. Figure 3.14(c) requires only 2 nodal-voltage equations when transformed into the form of Figure 3.14(d), and they can be written by inspection as

$$(G_1 + sC_1 + 1/sL_x + 1sLy)V_1 - V_2/sL_y = I_1 = E_1/R_1$$
$$(G_2 + sC_2 + 1/sL_z + 1sLy)V_2 - V_1/sL_y = 0$$

CHAPTER 4

Power transfer and allied concepts

INTRODUCTION

This chapter is concerned with the power absorbed by and transmitted through a network. Power is fundamental, for by definition it is the rate of doing work or dissipating energy, and is therefore the ultimate criterion of network performance in a system. For example, in a frequency-dependent system, the transmission boundaries are customarily taken as the frequencies at which the power has fallen to half its maximum value (the 3 dB frequencies; see Illustrations 4.5, 4.6, 4.8, and 4.9).

A linear, passive reciprocal network cannot deliver more power than it absorbs from a source, and the first consideration is therefore the conditions for maximum power transfer from a generator to such a network and from the network to a load. In the case of an active amplifying network the mechanism is different; for the output power is derived principally from a source separate from but controlled by the input signal, so that the output power may greatly exceed the available signal power. An amplifier, acting thus as a controlled generator but with linearity limits, delivers maximum possible power (without significant distortion) into a load whose ohmic value is an optimum quite different from that applicable to a passive network of unrestricted linearity. The distinction is high-lighted in Illustration 4.7.

The behaviour of a two-port network is realistically portrayed by comparing the load-power in the presence of the network with the power when the network is omitted and the load is directly connected to the generator. This is the practical basis of network specification in terms of its *insertion behaviour* (loss, gain, phase-change) in a given generator-load system. Allied concepts include *available power, transducer loss or gain, reflection loss* and *scattering* (reflection) *coefficients*.

A pure-reactance network loaded with a resistance is of special interest, for its transfer immittance (impedance or admittance) is easily correlated with the real part of its input immittance: this the very simple principle underlying Darlington's method of filter synthesis (see Illustrations 4.3, 4.15 and 4.16). While network synthesis may seem to be an an advanced topic, there is no reason why its simpler principles should not be appreciated at an early stage. See also Illustration 4.14, which relates insertion loss to natural frequencies (poles and zeros).

ILLUSTRATION 4.1

A generator of emf E_1 and series resistance R_1 is connected to a resistor $R_2 = kR_1$. Express the powers P_1 and P_2 dissipated in R_1 and R_2 respectively as functions of k, and show that P_2 is a maximum when $k = 1$. Plot curves of P_1, P_2, and the efficiency η, as functions of k.

Interpretation

$$P_1 = \frac{E_1^2}{R_1} \cdot \frac{1}{(1+k)^2} = P_{1\text{max}} \cdot \frac{1}{(1+k)^2}$$

$$P_2 = \frac{E_1^2}{R_1} \cdot \frac{k}{(1+k)^2} = P_{1\text{max}} \cdot \frac{k}{(1+k)^2}$$

P_2 is a maximum when

$$\frac{\mathrm{d}P_2}{\mathrm{d}k} = P_{1\text{max}} \frac{\mathrm{d}}{\mathrm{d}k}\left[\frac{k}{(1+k)^2}\right] = 0$$

whence

$$k = 1, \quad \text{or} \quad R_2 = R_1$$

The efficiency is

$$\eta = \frac{P_2}{P_1 + P_2} = \frac{k}{1+k}$$

P_1, P_2 and η are shown as functions of k in Figure 4.1.

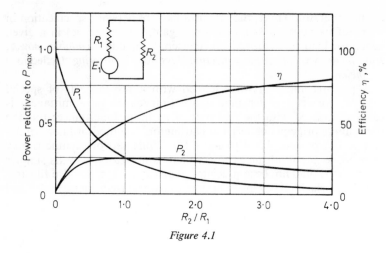

Figure 4.1

Comment

The wastage of 50% of the total power within the generator is inseparable from making the most of the power available when the generator has an intrinsic resistance. Figure 4.1 shows, however, that P_2 falls only slowly as k rises beyond the optimum value $k = 1$, whilst the efficiency rises initially at a greater rate. For example, when $k = 2$, or $R_2 = 2R_1$, P_2 falls from its maximum by only 11%, but η rises by 16.7%.

In a power system, involving megawatts, matching for a maximum power transfer is not only inconceivable from thermal considerations, but also is inconsistent with the role of the system and the need for high efficiency from economic considerations. An idealised system comprises, in principle, a resistanceless constant-voltage generator connected to paralleled loads, each adjusted for the abstraction of a desired power: there is not question of abstracting the maximum at any point.

In electronic systems the power available from a signal source (microphone, photocell, or other transducer or sensing device) is small, often in the μW range. Since noise power due to thermal agitation and other causes may not be negligible, even when the signal is amplified (amplifiers introduce additional noise), it is important to make the most of the small signal power at each junction of devices by matching for maximum power transfer. This is usually practicable, for the dissipation of 50% of the total power within the source is unlikely to present thermal difficulties when the power is small (although heat sinks are often necessary in the case of semiconductor devices). Matching often has also a functional role. For example, unless the impedance terminating a transmission line matches its characteristic impedance, the energy in a travelling wave is not wholly absorbed by the termination but is partially reflected from it.

ILLUSTRATION 4.2

(a) An emf generator has a complex series impedance $Z_1 = |Z_1| \underline{/\theta_1}$. Show that the power delivered to a load Z_2 is the maximum possible when $Z_2 = Z_1^*$, but that this condition is generally likely to be sharply frequency dependent.

(b) Show that if the conjugate condition is precluded, then a maximum of power (but not *the* maximum) is transferred to the load when $|Z_2| = |Z_1|$.

Interpretation

(a) Let $Z_1 = |Z_1| \underline{/\theta_1} = R_1 + jX_1$, and $Z_2 = |Z_2| \underline{/\theta_2} = R_2 + jX_2$. Then, when load and generator are connected, the power in the load Z_2 is

$$P_2 = |I|^2 R_2 = |E_1|^2 R_2 / [(R_1 + R_2)^2 + (X_1 + X_2)^2]$$

P_2 is a maximum when $X_1 + X_2 = 0$, or $X_2 = -X_1$, and the maximum possible when, additionally, $R_2 = R_1$ (see Illustration 4.1). The power absorbed by a load is thus the maximum possible when its impedance is the conjugate of the generator impedance, or symbolically, when

$$Z_2 = R_1 - jX_1 = |Z_1| \underline{/\theta_1} = Z_1^*$$

The component X of a complex impedance $Z = R + jX$ is always frequency dependent (consider the behaviour of the reactance elements L and C) while, except for a special case such as that in which the real component is attributable to a resistor directly in series with X, R is also frequency dependent. Thus, in general,

$$Z = Z(j\omega) = R(\omega) + jX(\omega)$$
$$Z^* = Z^*(j\omega) = R(\omega) - jX(\omega)$$

While a study of the parts $R(\omega)$, $X(\omega)$ of a generalised network impedance function is not simple, the sharp dependence of conjugate matching on frequency is easy to appreciate in terms of the simplest forms for $Z_1(j\omega)$ and $Z_2(j\omega)$. These are R_1 in series with $L_1 = X_1/\omega$ and R_2 in series with $C_2 = 1/X_2\omega$. Then, $Z_2(j\omega) = Z^*(j\omega)$ when $R_2 = R_1$ and and $X_2 = -X_1$. Let the resistances be assumed constant and let $\omega = \omega_0$ denote the frequency at which reactance cancellation occurs, so that

$$X_1(\omega_0) + X_2(\omega_0) = \omega_0 L_1 - 1/\omega_0 C_2 = 0.$$

Substituting $C_2 = 1/\omega_0^2 L_1$ then gives for any other frequency $\omega = \omega_0 + \delta\omega$,

$$X_1(\omega) + X_2(\omega) = \frac{L_1(\omega^2 - \omega_0^2)}{\omega} = \frac{L_1 \delta\omega(2\omega_0 + \delta\omega)}{\omega_0 + \delta\omega}$$

For $\delta\omega \ll \omega_0$ this gives

$$X_1(\omega) + X_2(\omega) \simeq 2\delta\omega L_1$$

which shows the net reactance to rise in proportion to twice the deviation in frequency from that at which the impedances are conjugates, which is the resonant frequency ω_0 for L_1 and C_2.

(b) Let the generator and load impedances be expressed in the form $Z_1 = |Z_1|/\underline{\theta_1}$ and $Z_2 = |Z_2|/\underline{\theta_2}$. Then

$$P_2 = \frac{|E_1^2 Z_2| \cos\theta_2}{(|Z_1 + Z_2|)^2}$$

$$= \frac{|E_1^2 Z_2| \cos\theta_2}{|Z_1|^2 + |Z_2|^2 + 2|Z_1 Z_2| \cos(\theta_1 - \theta_2)}$$

As $|Z_2|$ is varied, P_2 is a maximum when $\partial P_2/\partial |Z_2| = 0$. Performing this differentiation then shows this maximum to occur when $|Z_2| = |Z_1|$. It is not, however, the maximum possible power except when Z_1 and Z_2 are pure resistances, or when, additionally, the conjugate condition $\arg Z_2 = -\arg Z_1$ is satisfied.

Comment

When uniform behaviour over a wide band of frequencies is required, the ideal is for both generator and load to be purely resistive. While this condition can be approached closely in most parts of an audio-frequency system, a notable exception is in the connection of a loudspeaker to an amplifier.

The amplifier output impedance may be a constant resistance, but the loudspeaker has a complex impedance that is a function not only of the static impedance $R + j\omega L$ of its coil, but also a function of the mechanical parameters (mass, compliance, and damping) of the system in motion (motional impedance). Since annulment of the reactance would result in resonance and disproportionate sound output around the frequency of annulment, conjugate matching is precluded, and matching of the amplifier (or optimum loading, as appropriate) to the nominal modulus of the loudspeaker impedance must be tolerated. It is usually achieved with a wide-band audio-frequency transformer.

At radio frequencies, matching on a conjugate basis may be achieved with tuned circuits, and the departure from perfection over a fractionally small bandwidth is either tolerable, or can be made so by the use of band-pass circuits that preserve an approximately resistive maximum-power transfer condition over a specified frequency band.

ILLUSTRATION 4.3

A two-port network composed solely of inductors and capacitors that may be assumed pure is interposed between a sinusoidal (or cosinusoidal) generator and a load. Justify the fact that the mean power delivered to the load is equal to the power absorbed by the real part of the input impedance or admittance of the loaded network.

A sinusoidal generator having an angular frequency $\omega = 10^6$ rad/s and a purely resistive internal impedance of $1000 \, \Omega$ is to be matched for maximum power transfer to a load comprising a resistor of $100 \, \Omega$ in series with an inductor of $100 \, \mu H$. The matching circuit is an L-network in which the shunt arm, on the input-side, is a capacitor C; and the series arm, on the load-side, is an inductor L. Calculate the required values for L and C.

Interpretation

(a) The rate of energy supply or power input to a device is expressed by

$$P = \frac{1}{T} \int_0^T w \, dt$$

Let the voltage across a pure inductor under steady-state conditions be $v = \hat{V} \cos \omega t = \hat{V} \sin (\omega t + \pi/2)$. Then,

$$i = \frac{1}{L} \int_0^t v \, dt = \frac{\hat{V}}{\omega L} \sin \omega t$$

and the mean power, or rate of energy supply averaged over the cycle, is

$$P = \frac{\omega}{2\pi} \int_{t=0}^{t=2\pi/\omega} vi \, dt = - \frac{\hat{V}^2}{2\pi L} \left[\frac{1}{4\omega} \cos 2\omega t \right]_{t=0}^{t=2\pi/\omega} = 0$$

It follows by duality that the mean power absorbed by a pure capacitor is also zero. Hence, in rms phasor notation, the power $P_2 = |V_2| \cdot |I_2| \cdot \cos \phi_2$ transmitted to a load Z_2 through a pure $L-C$ network is equal to the power $|V_1| \cdot |I_1| \cdot \cos \phi_1$ dissipated in the input impedance Z_1 of the terminated network; or,

$$|V_1| \cdot |I_1| \cdot \cos \phi_1 = |V_2| \cdot |I_2| \cdot \cos \phi_2$$

$$|I_1|^2 \cdot |Z_1| \cdot \cos \phi_1 = |I_2|^2 \cdot |Z_2| \cdot \cos \phi_2$$

$$|I_1|^2 R_1 = |I_2|^2 R_2$$

and similarly,

$$|V_1|^2 G_1 = |V_2|^2 G_2$$

(b) Since the matching network is purely reactive the mean power entering it when terminated by the load is equal to the mean power dissipated in the load. Thus, it is sufficient to satisfy maximum power transfer conditions on one side of the network only, for they are then automatically satisfied on the other.

Let R_2, L_2 denote the load parameters, and let R_1 be the generator resistance. From the input terminals the network and the load appear as a parallel circuit, comprising C in parallel with the series elements L, L_2 and R_2. The input admittance is therefore immediately

$$Y_{in} = j\omega C + G - jB$$

$$= j\omega C + \frac{R_2}{R_2^2 + \omega^2 (L + L_2)^2} - j \frac{\omega(L + L_2)}{R_2^2 + \omega^2 (L + L_2)^2}$$

Assuming first that the total susceptance is zero, the required condition for maximum power transfer is

$$\frac{1}{R_1} = \frac{R_2}{R_2^2 + \omega^2 (L + L_2)^2}$$

whence, on substituting values

$$L = 200 \ \mu H$$

The susceptance is zero, as initially assumed, when

$$\omega C - \frac{\omega(L + L_2)}{R_2^2 + \omega^2 (L + L_2)^2} = 0$$

from which

$$C = 3000 \ pF$$

Comment

The load impedance is

$$Z_L = R_2 + j\omega L_2 = 100 + j100$$

The output impedance of the L-network closed on the input side by the generator resistance R_1 is

$$Z_{\text{out}} = j\omega L + \frac{1}{R_1^{-1} + j\omega C}$$

$$= j200 + \frac{1}{10^{-3}(1+j3)}$$

$$= j200 + \frac{10^3(1-j3)}{10}$$

$$= j200 + 100 - j300$$

$$= Z_L^*, \text{ the conjugate of } Z_L$$

This confirms the simultaneous satisfaction of the maximum power transfer conditions at both ports of the matching network.

ILLUSTRATION 4.4

Identify the concept of an ideal transformer with the postulate of an hypothetical two-port network that neither stores nor dissipates energy.

Interpretation

(a) Let v_1, i_1 and v_2, i_2 denote respectively the input and output voltages and currents when a load is connected to the output terminals of the postulated network. Then, since this is supposed neither to store nor to dissipate energy,

$$\int_0^t v_1 i_1 \, dt = \int_0^t v_2 i_2 \, dt$$

for all values of t.
Differentiating both sides,

$$v_1 i_1 = v_2 i_2$$

Let $v_1/v_2 = \lambda$. Then $i_1/i_2 = 1/\lambda$, and if an impedance of transient form v/i is denoted by $Z(p)$,

$$\frac{v_1}{i_1} = Z_1(p) = \lambda v_2 \cdot \frac{\lambda}{i_2} = \lambda^2 v_2/i_2 = \lambda^2 Z_2(p)$$

Thus, the pure transformation of a voltage in a real ratio λ, to the exclusion of energy storage and dissipation, is accompanied by current and impedance transformations that are in the exact ratios $1/\lambda$ and λ^2, respectively.

(b) Consider the transformer of Figure 4.2, in which, as a first step in idealisation the inductive elements are assumed to be non-dissipative. The input voltage is expressed as a function of the general complex-variable s, so that the analysis may embrace the time domain, and permit the impedance concept to be used in its widest sense.

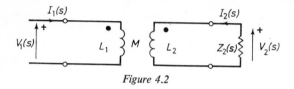

Figure 4.2

For the voltage and current orientations chosen,

$$Z_{11}(s)I_1(s)+Z_{12}(s)I_2(s) = V_1(s)$$
$$Z_{21}(s)I_1(s)+Z_{22}(s)I_2(s) = V_2(s) = -Z_2(s)I_2(s)$$

where

$$Z_{11}(s) = sL_1, \quad Z_{22}(s) = sL_2, \quad Z_{12}(s) = Z_{21}(s) = sM = sk\sqrt{(L_1L_2)}$$

From the second equation,

$$\frac{I_2(s)}{I_1(s)} = \frac{-Z_{21}(s)}{Z_{22}(s)+Z_2(s)} = -\frac{sk\sqrt{(L_1L_2)}}{sL_2+Z_2(s)}$$

Then, substituting for either current in terms of the other and dividing the equations,

$$\frac{V_1(s)}{V_2(s)} = \frac{s^2L_1L_2(1-k^2)+sL_1Z_2(s)}{sk\sqrt{(L_1L_2)}\cdot Z_2(s)}$$

As $k \to 1$, in the limit

$$\frac{V_1(s)}{V_2(s)} = \sqrt{\frac{L_1}{L_2}}$$

and

$$\frac{I_1(s)}{I_2(s)} = -\frac{sL_2+Z_2(s)}{s\sqrt{(L_1L_2)}} = -\frac{1+Z_2(s)/sL_2}{\sqrt{(L_1/L_2)}}$$

For s and $Z_2(s)$ finite, $Z_2(s)/sL_2 \to 0$ as $L_2 \to \infty$. Then,

$$\frac{I_1(s)}{I_2(s)} \to -\sqrt{\frac{L_2}{L_1}}$$

Since indefinitely great values for L_1 and L_2 are incompatible with physical realism, yet mathematical realism remains in their fixed ratio, it is now preferable to denote $\sqrt{(L_1/L_2)}$ by a real positive constant λ. Thus, in the limit, as $L_2 \to \infty$ and $L_1 \to \infty$,

$$\frac{V_1(s)}{V_2(s)} = \lambda$$

$$\frac{I_1(s)}{I_2(s)} = -\frac{1}{\lambda}$$

and

$$\frac{V_1(s)}{I_1(s)} = Z_1(s) = \lambda V_2(s) \cdot \frac{\lambda}{-I_2(s)} = \lambda^2 Z_2(s)$$

or

$$\frac{Z_1(s)}{Z_2(s)} = \lambda^2$$

Comment

The results obtained through the idealisation of a physical transformer are identical in form with those obtained from energy considerations without reference to any particular structure. The negative current ratio means that the actual direction of $I_2(s)$ is opposite to that assumed for the purpose of setting-up the equations. The convention used is, however, immaterial, so long as the Kirchhoff-law equations are correctly correlated with it: if $I_2(s)$ had been taken as clockwise, and $V_2(s)$ as the voltage fall in the sense of $I_2(s)$, the result $I_1(s)/I_2(s) = 1/\lambda$ would have been obtained; i.e., the directions and polarities assumed would then, by chance, have coincided with the true ones. As λ is wholly real and independent of s, the transformation properties of the ideal transformer are independent of both time and frequency. As a passive, reciprocal inference from the physical transformer, the ideal transformer might be defined as a hypothetical two-port device that neither stores nor dissipates energy, and which embodies the attributes of a perfect but unrealisable two-winding transformer.

There are many other kinds of two-port convertor, realisable with active circuitry (transistors), that have immittance-transforming properties quite different from those of the passive transformer. An example is the *gyrator*, for which the admittance matrix is ideally

$$[Y] = \begin{bmatrix} 0 & g \\ -g & 0 \end{bmatrix}$$

When terminated with an admittance Y, the gyrator input admittance is $Y_{in} = g^2/Y$, so that if Y is the admittance sC of a capacitor, Y_{in} becomes the admittance of an inductor $L = C/g^2$. The gyrator thus realises an inductor from a capacitor.

ILLUSTRATION 4.5

(a) A sinusoidal emf generator represented by E_0 in series with R_0 is connected to a resistor R in series with a capacitor C. If P is the power in R at frequencies above which $1/\omega C$ is negligible, show that the power in R is $P/2$ when the frequency is lowered to $\omega = \omega_1 = 1/C(R_0+R)$, and that at this half-power frequency, the voltage V across R is attenuated by 3 db and advanced in phase by 45°, relative to its limiting value at high frequencies.

(b) The same generator is connected instead to a resistor R in parallel with a capacitor C. Show that there is now an upper half-power frequency, above the range for which ωC is negligible, given by $\omega_2 = (G_0+G)/C$, where $G_0 = 1/R_0$ and $G = 1/R$; and show that the voltage V across R is now retarded in phase by 45°, relative to its limiting value at low frequencies.

Interpretation

(a) For frequencies beyond which $1/\omega C$ is negligible, the net series impedance is the minimum value, and therefore the power in R is the maximum, P. As the frequency is reduced and $1/\omega C$ becomes significant, the power in R is halved at the frequency for which the current magnitude is divided by $\sqrt{2}$. This occurs when $|Z| = \sqrt{2}(R_0+R)$, and by reference to a triangle representing $Z = R_0+R-j/\omega C$, $|Z| = \sqrt{2}(R_0+R)$ when $R_0+R = 1/\omega C$ and the triangle has the proportions $1 : 1 : \sqrt{2}$. Additionally from this triangle, arg $Z = -45°$. The half-power frequency is therefore

$$\omega = \omega_1 = 1/C(R_0+R)$$

and at this frequency V is leading by 45° on E_0, and on its high frequency value which is in-phase with E_0, for

$$V = \frac{E_0R}{|Z|/\underline{45°}} = \frac{E_0R}{\sqrt{2}(R_0+R)}\underline{/45°}$$

The power in R is reduced in the ratio $2 : 1$ relative to its maximum value P at high frequencies, and the attenuation in decibels is therefore $10 \log_{10} 2 = 3{\cdot}01$ db. Alternatively, the voltage V is reduced in the ratio $\sqrt{2} : 1$, and the attenuation is equivalently given by $20 \log_{10} \sqrt{2} = 3{\cdot}01$ db.

(b) A completely parallel circuit may be formed by transforming the generator into a current source E_0/R_0 in parallel with $G_0 = 1/R_0$. This circuit has a high-frequency half-power of 3 db frequency when $|Y| = \sqrt{2}(G_0+G)$ and the magnitude of the voltage V across G is reduced in the ratio $\sqrt{2}:1$ relative to its limiting value at low frequencies. By reasoning similarly to case (a), $|Y| = \sqrt{2}(G_0+G)$ when $G_0+G = \omega C$ and arg $Y = 45°$. Thus,

$$\omega = \omega_2 = (G_0+G)/C$$

and V lags by 45° on its low-frequency value which is in-phase with E_0, for

$$V = \frac{E_0/R_0}{Y\underline{/45°}} = \frac{E_0G_0}{\sqrt{2}(G_0+G)}\underline{/-45°}$$

Comment

The basic circuits considered are very important in the context of electronic amplifiers. Circuit (a) corresponds to the equivalent circuit commonly used to study the response of an untuned capacitance-coupled amplifier at low frequencies at which the coupling capacitor is a controlling factor, while circuit (b) has the form of the high-frequency equivalent circuit, in which the shunt capacitance due to strays and the input of the following stage is increasingly significant with rising frequency.

The principle of establishing half-power or 3 db frequencies by equating net reactance to resistance (series arrangements) or net susceptance to conductance (parallel arrangements) is quite general and is applicable also to tuned circuits. The 3 db frequencies are commonly taken as effective cut-off frequencies in the response of a network or amplifier.

ILLUSTRATION 4.6

Show that Figure 4.3(b) is an exact equivalent representat on of Figure 4.3(a), where k is the coefficient of coupling between the inductors L_1 and L_2 (which may be assumed pure), and T is an ideal transformer of impedance transformation ratio $k^2L_1 : L_2$ in the sense indicated. Justify briefly the approximations to Figure 4.3(b) that are usually made in the case of a practical audio-frequency transformer.

A transistor used as a power amplifier in the common-emitter circuit is coupled to a resistance load of $5\,\Omega$ with an audio-frequency transformer having a primary inductance of 5 H. For an input signal $i_b = 200 \sin \omega t\ \mu A$, the collector voltage and current vary about their

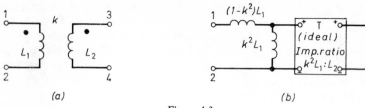

Figure 4.3

quiescent values of 16 V and 19.2 mA from 5·5 to 26·5 V and 8·0 to 29·0 mA, respectively, when the frequency is in the middle of the audio range.

(a) What is the effective mid-frequency ac load impedance between collector and emitter, and what, with reasonable assumptions, is the secondary inductance of the transformer?

(b) What is the power in the load at mid-frequency, and what is the lower half-power (3 db) frequency?

(c) What is the power gain of the stage if the transistor has an input resistance of $1000\,\Omega$?

(P.C.L., BSc. (Hons.), C.N.A.A., Electronics, Year 2)

Interpretation

(1) The equivalence may be confirmed by equating three independent sets of impedances at corresponding terminal-pairs under like conditions at the other pairs. The simplest valid conditions are two under open-circuit and one under short-circuit. Noting that the input impedance of the ideal transformer T is infinite under open-circuit and zero under short-circuit conditions, Figure 4.3(b) gives

$$3\text{–}4 \text{ O.C.:} \quad Z^0_{1-2} = j\omega L_1(1-k^2) + j\omega k^2 L_1 = j\omega L_1$$

$$1\text{–}2 \text{ O.C.:} \quad Z^0_{3-4} = j\omega k^2 L_1 \times \frac{L_2}{k^2 L_1} = j\omega L_2$$

$$3\text{–}4 \text{ S.C.:} \quad Z^s_{1-2} = j\omega L_1(1-k^2) + 0 = j\omega L_1(1-k^2)$$

and Figure 4.3(a) gives

$$3\text{–}4 \text{ O.C.:} \quad Z^0_{1-2} = j\omega L_1 \quad \text{(by inspection)}$$

$$1\text{–}2 \text{ O.C.:} \quad Z^0_{3-4} = j\omega L_2 \quad \text{(by inspection)}$$

$$3\text{–}4 \text{ S.C.:} \quad Z^s_{1-2} = j\omega L_1 + \omega^2 M^2/j\omega L_2$$

$$= j\omega L_1 + \omega^2 k^2 L_1 L_2/j\omega L_2$$

$$= j\omega L_1(1-k^2)$$

The identity of corresponding impedances confirms that Figure 4.3(a) and (b) are equivalent.

For a good iron-cored transformer, k is close to unity and the self inductances are closely proportional to the square of the winding turns. Thus $L_1(1-k^2)$ is small, the shunt inductance k^2L_1 is approximated by L_1, and the ideal transformer T is assumed to have an impedance transformation ratio $L_1 : L_2$ or $N_1^2 : N_2^2$. The transformer performance is impaired by the shunt inductance L_1 at low frequencies, and by the series leakage inductance $L_1(1-k^2)$ at high frequencies. At intermediate frequencies the performance is close to that of the ideal transformer T.

(2) The numerical problem

(a) The effective collector-emitter ac load at mid-frequency is

$$R = \frac{\Delta V}{\Delta I} = \frac{26 \cdot 5 - 5 \cdot 5}{(29-8)10^{-3}} = 1000 \, \Omega$$

The transformer performance at mid-frequency is approximately ideal, so that the secondary inductance must be

$$L_s = L_p \times \frac{5}{1,000} = 25 \times 10^{-3} \, \text{H}$$

(b) The voltage and current oscillate with peak values of $\Delta V/2$ and $\Delta I/2$ about the given quiescent values, and the mid-frequency output power into the 5-Ω load is therefore

$$P_{\text{out}} = \frac{10 \cdot 5}{\sqrt{2}} \times \frac{10 \cdot 5 \times 10^{-3}}{\sqrt{2}} = 55 \, \text{mW}$$

Figure 4.4(a) represents the transformer at low frequencies. The ideal transformer T has an impedance ratio of 1000 : 5, in accordance with the assumed approximation to ideal performance at an inter-

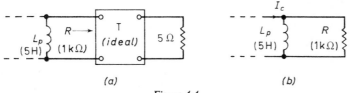

(a)   (b)

Figure 4.4

mediate or mid-frequency, at which its input impedance is known to be a resistance $R = 1000 \, \Omega$ from the voltage and current swings (see (a)). The effective transistor load may therefore be simplified to the form of Figure 4.4(b). As the collector-emitter resistance of the transistor is very much greater than $1000\,\Omega$, its shunting effect may be ignored and the lower 3 db frequency is then given closely by equating the susceptance $1/\omega L_p$ to the conductance $1/R$ in Figure 4.4(b). This gives

$$f_1 = \frac{\omega_1}{2\pi} = \frac{R}{2\pi L_p} = \frac{1000}{6\cdot28\times5} = 32\,\text{Hz}$$

(c) The input power is

$$P_{in} = \left[\frac{200}{\sqrt{2}}\times10^{-6}\right]^2\times10^3 = 2\times10^{-2}\,\text{mW}$$

From (a) the output power P_{out} is 55 mW, so that the power gain is

$$P_{out}/P_{in} = 55/(2\times10^{-2}) = 2750$$

Comment

(1) An alternative to the equivalence shown in Figure 4.3(b) is a circuit comprising, from left to right, an ideal transformer of impedance ratio $L_1 : k^2 L_2$, a shunt inductor $k^2 L_2$, and a series inductor $L_2(1-k^2)$.

(2) High-frequency performance is more complex than low, for the transformer is complicated by self and inter-winding capacitances that result in complex resonance effects. When the transformer is grounded on both sides, the capacitance network reduces to the primary and secondary self-capacitances C_p and C_s and an inter-winding capacitance C_{ps}. Ignoring C_{ps}, C_p may be referred to the secondary as $C_p(N_p/N_s)^2$ so that a resonance is predictable when $\omega L_2(1-k^2) = 1/\omega[C_s + C_p(N_p/N_s)^2]$. Beyond this resonance the response falls rapidly.

ILLUSTRATION 4.7

In the case of a linear, reciprocal two-port network of initially passive elements, the power delivered to a resistance load is the maximum possible for any given excitation when the load resistance equals the output resistance of the network. But in the case of a device of restricted linearity, such as an electronic amplifier, there is an optimum value of load resistance that does not correspond to the output resistance of the device. Distinguish between the two cases.

Interpretation

(a) Let the open-circuit output voltage of the linear reciprocal network be V_0 when an excitation V_1 is applied to the input terminals, and let R_0 be the output resistance when V_1 is reduced to zero. Then the Thevenin-type equivalent generator comprises R_0 in series with V_0, where $V_0 = \lambda V_1$ and both R_0 and λ are functions of the network elements alone and are independent of V_1.

The power in a load R_L is

$$P_L = \frac{V_0^2 R_L}{(R_0 + R_L)^2} = V_1^2 \cdot \frac{\lambda^2 R_L}{(R_0 + R_L)^2}$$

and for R_L as variable while λ and R_0 are constants, P_L is a maximum for any value of V_1 when $R_L = R_0$ (see Illustration 4.1).

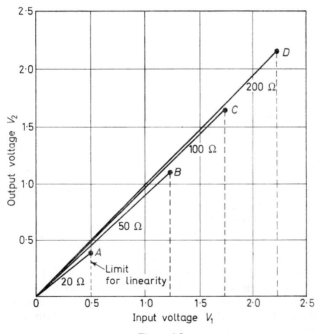

Figure 4.5

(b) Figure 4.5 shows the voltage-transfer characteristics for a transistor in the common-collector (emitter-follower) circuit, for a range of load resistances. For each resistance, the transfer characteristic is linear only for the input voltages up to the nominal limits indicated by the points *A—D*. For freedom from distortion, the power in any load resistance, V_2^2/R_L, is therefore restricted to a maximum set by the maximum permissible input voltage V_1 for that load. Since the permissible input and output voltages are themselves functions of the load resistance, there exists a unique combination of load resistance and input voltage for which the undistorted load-power is the maximum possible. This resistance is referred to as the *optimum load*.

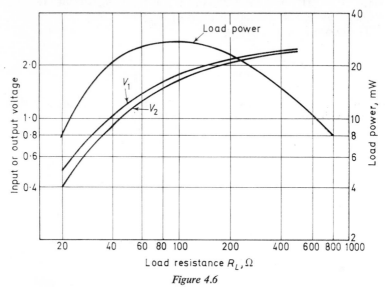

Figure 4.6

Figure 4.6 shows the maximum permissible values of V_1 and V_2 corresponding to the points *A—D* in Figure 4.5, and the attendant maximum power in R_L, plotted as functions of R_L. The maximum possible power is estimated to coincide with $R_L = 100\,\Omega$, $V_1 = 1\cdot72$ V $V_2 = 1\cdot65$ V and a voltage transfer ratio ('gain') of $V_2/V_1 = 0\cdot96$. The optimum load is therefore about $100\,\Omega$; and with this, the maximum of undistorted power will be dissipated for all values of V_1 up to the maximum permissible value of $1\cdot72$ V. By contrast, the effective output resistance of the circuit, as a determined with V_1 at a low value well within the range of linearity, is about $8\,\Omega$. This was estimated experimentally by adjusting the load, with V_2 monitored with a CRO for waveform purity, until V_2 fell to $\frac{1}{2}$ its open-circuit value.

Comment

The distinction between the two cases is bound up with the distinction between a system that is not limited by excitation and one that is. In the case of a truly linear system having no overload point (strictly hypothetical), the load-power is proportioned to the square of the rms value of the excitation for all values of excitation and load; and therefore the absolute value of the excitation is irrelevant to a determination of that load which will absorb the maximum of power attainable with any excitation. But when the system has restricted linearity and an overload point that is itself a function of the load, as in the case of an electronic amplifier, the load for maximum possible undistorted power may be very different from the intrinsic output resistance of the device under small excitation conditions; for now both permissible excitation and load must be considered conjointly.

ILLUSTRATION 4.8

A series-tuned circuit comprising a capacitor C in series with an inductor L of series resistance R is connected to a sinusoidal generator of emf E and series resistor r. Show that the current at any frequency f may be expressed in the form

$$I = \frac{I_r}{1+jQ_r \left[\dfrac{f}{f_r} - \dfrac{f_r}{f} \right]}$$

where I_r is the current at the resonant frequency f_r and $Q_r = \omega_r L/(R+r) = 1/\omega_r C(R+r)$. Thence show that the fractional bandwidth is

$$(f_2-f_1)/f_r = 1/Q_r$$

when f_1 and f_2 are the half-power frequencies below and above f_r.

Interpretation

$$I = \frac{E}{R+r+j(\omega L-1/\omega C)}$$

$$= \frac{E/(R+r)}{1+j\left[\omega \cdot \dfrac{L}{R+r} - \dfrac{1}{\omega} \cdot \dfrac{1}{C(R+r)} \right]}$$

But $L/(R+r) = Q_r/\omega_r$ and $1/C(R+r) = Q_r\omega_r$, and therefore

$$I = \frac{E(R+r)}{1+jQ_r \left[\dfrac{\omega}{\omega_r} - \dfrac{\omega_r}{\omega} \right]} = \frac{I_r}{1+jQ_r \left[\dfrac{f}{f_r} - \dfrac{f_r}{f} \right]}$$

At the half-power (3 db) frequencies f_1 and f_2, $|I| = |I_r|/\sqrt{2}$. This is so when

$$Q_r\left[\frac{f_r}{f_1} - \frac{f_1}{f_r}\right]_{f_1 < f_r} = 1$$

and

$$Q_r\left[\frac{f_2}{f_r} - \frac{f_r}{f_2}\right]_{f_2 > f_r} = 1$$

whence

$$f_r^2 = f_1 f_2$$

Then at $f = f_2$, for example,

$$\frac{f_2}{f_r} - \frac{f_r}{f_2} = \frac{f_2^2 - f_r^2}{f_r f_2} = \frac{1}{Q_r}$$

whence, on writing $f_r^2 = f_1 f_2$,

$$\frac{f_2 - f_1}{f_r} = \frac{1}{Q_r}$$

Comment

The result, $(f_2 - f_1)/f_r = 1/Q_r$, is exact for $R + r$ constant, which is likely to be a valid assumption when the ratio $f_2 : f_1$ is small. The expression provides a simple criterion for the selectivity of the circuit in terms of its effective Q-factor; or alternatively, it may be used to measure Q by observing the frequency deviations about f_r necessary to reduce the current by the factor $1/\sqrt{2}$.

ILLUSTRATION 4.9

A parallel-tuned circuit comprises an inductor of series components L and R, in parallel with a pure capacitor C. Express the admittance of the circuit in terms of the Q-factor of the inductor, and derive expressions for the pure components L_0 and G which, in parallel with one another and with C, give a circuit exactly equivalent to the given one at an angular frequency ω. Justify briefly the simplifications possible in practice through treating high-Q amplifier problems by duality with relationships applying to series-tuned circuits.

A common-base transistor amplifier, whose output circuit may be approximately represented by a constant current $0.98I_e$ in parallel with a conductance of 2×10^{-6} S, and whose input impedance has a magnitude of $50\,\Omega$, has a parallel-tuned circuit of inductance $500\ \mu H$ and Q-factor 200 in parallel with a capacitor of $500\ \mu\mu F$ as its collector-base load. Calculate the voltage gain V_{cb}/V_{eb} at the parallel-resonant frequency f_r, and approximately at $1 \cdot 05 f_r$.

(P.C.L., BSc. (Hons), C.N.N.A., Electronics, Year 2).

Interpretation

(a) For an impedance $Z = R + jX$,

$$Y = \frac{1}{Z} = \frac{R}{|Z|^2} - j\frac{X}{|Z|^2} = G - jB$$

From this general result the admittance of the parallel-tuned circuit may be quoted as

$$Y = \frac{R}{R^2 + \omega^2 L^2} - j\frac{\omega L}{R^2 + \omega^2 L^2} + j\omega C$$

Then substituting $R = \omega L/Q$ gives

$$Y = \frac{Q}{\omega L(1 + Q^2)} - j\frac{Q^2}{\omega L(1 + Q^2)} + j\omega C$$

The admittance of a conductance G in parallel with a pure inductor L_0 and the capacitor C is given directly by

$$Y_0 = G + j\omega C - j/\omega L_0$$

This circuit is exactly equivalent to the given parallel-tuned circuit when $Y_0 = Y$. This is so when

$$G = Q/\omega L(1 + Q^2)$$

and

$$1/\omega L_0 = Q^2/\omega L(1 + Q^2)$$

or

$$L_0 = L(1 + 1/Q^2)$$

For Q as low as 10, for example, the error is only 1% if L_0 is approximated by L and G by $1/Q\omega L$. Hence for reasonable Q values, the admittance of the parallel-tuned circuit is closely given by

$$Y = G + j\omega C - j/\omega L$$

which is the dual of

$$Z = R + j\omega L - j/\omega C$$

for a series circuit, except that whereas R is substantially constant, G is frequency dependent. But G varies less rapidly with frequency than the susceptance $\omega C - 1/\omega L$, and from the parallel-resonant frequency f_r to either of the half-power (3 db) frequencies, the change is quite small ($\delta G \simeq G_r/2Q_r$, where $G_r = 1/Q_r\omega_r L$). Moreover, this change is likely to be masked by constant conductance contributed by an amplifier or other source, so that G may usually be treated as constant and equal to $1/Q_r\omega_r L$, where Q_r is the inductor Q-factor at ω_r. The voltage in the neighbourhood of f_r in response to a current source is then the dual of the current in a series circuit in response to an emf source, and the simple relations for this apply by duality as close approximations. Thus, the voltage V across a parallel-tuned circuit in response to a parallel-connected current source I_s with shunt conductance G_s is

$$V = \frac{I_s}{Y} = \frac{I_s}{G_s + G + j(\omega C - 1/\omega L)}$$

and by duality with the series circuit (see Illustration 4.8),

$$V = \frac{V_r}{1 + jQ_r \left[\dfrac{f}{f_r} - \dfrac{f_r}{f} \right]}$$

as a close approximation.

(b) *The problem*

Let G_0 and I_0 denote the parameters of the current generator representing the output circuit of the simplified transistor. As Q is high, the components L and G in the equivalent parallel representation for the tuned circuit are given with negligible error by

$$L = 500 \ \mu\mathrm{H}$$

$$G = \frac{1}{Q\omega_r L} = \frac{1}{Q}\sqrt{\frac{C}{L}} = 5 \times 10^{-6} \mathrm{S}$$

At $f = f_r$, the admittance of the tuned circuit reduces to G, and

$$V_{cb} = I_0/(G_0 + G) = 0.98I_e/7 \times 10^{-6}$$

The input voltage is $V_{eb} = R_{in}I_e = 50I_e$ and the voltage gain at f_r is therefore

$$A_r = \frac{V_{cb}}{V_{eb}} = \frac{0.98I_e}{7 \times 10^{-6}} \cdot \frac{1}{50I_e} = 2.8 \times 10^3$$

The effective Q factor of the tuned circuit in the presence of the shunt conductance G_0 is

$$Q_{\text{eff}} = \frac{1}{(G_0+G)\,\omega_r L} = \frac{1}{G_0+G}\cdot\sqrt{\frac{C}{L}} = 143$$

Then, by duality with the response formula for the series circuit, the gain at $f = 1\cdot05 f_r$ is

$$|A| = \left|\frac{A_r}{1+jQ_{\text{eff}}\left[\dfrac{f}{f_r}-\dfrac{f_r}{f}\right]}\right|$$

$$= \frac{A_r}{\sqrt{1+\left\{Q_{\text{eff}}\left[\dfrac{f}{f_r}-\dfrac{f_r}{f}\right]\right\}^2}} = 215$$

Comment

The physical dissymmetry in the distribution of the parameters L, R, C in a parallel-tuned circuit may make exact calculations tedious, but for most purposes, where interest is confined to the significant range of its response, representation by physically parallel components G, L, C is valid as a close approximation as demonstrated, and simplifies calculations by reducing them to the duals of those applying to a series-tuned circuit.

It should be noted that the simple equivalent representation used here for the transistor is a rough approximation only, for it ignores reverse transmission and also the reactive elements necessary to simulate high-frequency behaviour. However, it might be regarded as a fair representation for a unilateralised transistor, in which reverse transmission (Y_{12} or h_{12}) has been cancelled by the addition of suitable external circuitry.

ILLUSTRATION 4.10

Define the term *image impedance* as applied to a linear passive reciprocal two-port network, and show that at either of the ports it is the geometric mean of the impedances under open and short-circuit conditions at the other.

Interpretation

Consider the input impedance Z_{in} of the network when its output port is closed by an impedance Z_2, and its output impedance Z_{out} when its input port is closed by an impedance Z_1. There are unique values of Z_1 and Z_2 such that $Z_{\text{in}} = Z_1$ and $Z_{\text{out}} = Z_2$ when both ports are closed simultaneously by these values: at each port the impedance looking into the port is the *image* of the impedance closing it. The notations Z_{i1} and Z_{i2} commonly denote these image impedances.

Let the network be defined, for the usual 2-port convention, by the equations

$$V_1 = AV_2 - BI_2$$
$$I_1 = CV_2 - DI_2 \tag{1}$$

and

$$V_2 = DV_1 - BI_1$$
$$I_2 = CV_1 - AI_1 \tag{2}$$

The input impedance V_1/I_1 is to be Z_{i1} when the output port is closed by Z_{i2} and $V_2 = -Z_{i2}I_2$, while the output impedance V_2/I_2 is to be Z_{i2} when the input port is closed by Z_{i1} and $V_1 = -Z_{i1}I_1$. Imposing these conditions on equations (1) and (2),

$$Z_{i1} = (AZ_{i2} + B)/(CZ_{i2} + D) \tag{3}$$
$$Z_{i2} = (DZ_{i1} + B)/(CZ_{i1} + A) \tag{4}$$

whence $DZ_{i1} = AZ_{i2}$. Substituting in equation (3) for Z_{i2} and equation (4) for Z_{i1} gives

$$Z_{i1} = \sqrt{\frac{AB}{CD}} \quad \text{and} \quad Z_{i2} = \sqrt{\frac{DB}{CA}} \tag{5}$$

But by reference to equations (1) or (2),

$$\frac{A}{C} = \frac{V_1}{I_1}\bigg|_{I_2=0} = Z_{\text{in}}^0, \qquad \frac{B}{D} = \frac{V_1}{I_1}\bigg|_{V_2=0} = Z_{\text{in}}^s$$

$$\frac{D}{C} = \frac{V_2}{I_2}\bigg|_{I_1=0} = Z_{\text{out}}^0, \qquad \frac{B}{A} = \frac{V_2}{I_2}\bigg|_{V_1=0} = Z_{\text{out}}^s$$

Hence

$$Z_{i1} = \sqrt{(Z_{\text{in}}^0 Z_{\text{in}}^s)}, \quad Z_{i2} = \sqrt{(Z_{\text{out}}^0 Z_{\text{out}}^s)}, \tag{6}$$

Comment

The image impedances govern power transfer at both ports of a network. When they are wholly real, maximum power transfer from a resistive generator to the network, and from the network to a resistive load, is attainable.

An allied concept is the *image transfer coefficient* $\Gamma = \alpha + j\beta = \frac{1}{2}\log_e(V_2 I_2 / V_1 I_1)$, defined under the condition of image impedance termination at the output port. When a network is operated between its image impedances, the attenuation and phase-change are exactly expressed by α (in nepers) and β (in radians). The symmetrical type of network is a special case for which the image impedances are equal and may be called the *characteristic impedance*, while Γ becomes the *propagation coefficient* $\gamma = \alpha + j\beta = \log_e I_2 / I_1 = \log_e V_2 / V_1$, for the condition of characteristic impedance termination. The concepts Z_i and Γ form the basis of the *image-parameter approach* to transmission networks.

ILLUSTRATION 4.11

Distinguish between the *insertion loss* or gain and the *transducer loss* or gain of a linear two-port network interposed between a generator of series resistance R_1 and emf E_1, and a load resistor R_2. Find the transducer power ratio for an ideal transformer in terms of its voltage ratio λ, and thence obtain a formula for the loss in dB that would be incurred if load and generator were directly connected instead of being matched with a non-dissipative network.

Interpretation

The *insertion loss or gain* of a linear two-port network is the attenuation or gain in load power occasioned by inserting the network between a given generator and load. Let the load currents and voltages be I_2, V_2 with the network present, and I_{20}, V_{20} when the load is connected directly to the generator. The *insertion power ratio* is

$$\frac{P_2}{P_{20}} = \frac{|I_2|^2 R_2}{|I_{20}|^2 R_2} = \left|\frac{I_2}{I_{20}}\right|^2 = \left|\frac{V_2}{V_{20}}\right|^2 \tag{1}$$

and the *insertion phase-change* is the argument of the current or voltage ratio. When $P_2 < P_{20}$, the insertion loss in dB is expressed by

$$\begin{aligned}
L_{\text{insn}} &= 10\log_{10}(P_{20}/P_2) \\
&= 20\log_{10}|I_{20}/I_2| = 20\log_{10}|V_{20}/V_2|, \quad \text{dB} \tag{2}
\end{aligned}$$

The *transducer loss or gain* is the attenuation or gain in load power caused by the network, but relative to the maximum power the generator could deliver if, in general, it were conjugate-matched to the load. This *available power P_{av}* is $|E_1|^2/4R_1$, and the transducer power ratio is therefore

$$\frac{P_2}{P_{av}} = \frac{P_2}{P_{20}} \cdot \frac{P_{20}}{P_{av}} = \left|\frac{I_2}{I_{20}}\right|^2 \cdot \frac{4R_1R_2}{(R_1+R_2)^2} \tag{3}$$

which can be interpreted as loss or gain in dB as before.

Applying Thevenin's theorem to the generator E_1, R_1 followed by the ideal transformer of voltage-ratio λ gives an equivalent generator of emf λE_1 and resistive impedance $\lambda^2 R_1$. The current this produces in R_2 is $I_2 = \lambda E_1/(R_2 + \lambda^2 R_1)$ and the transducer power ratio is therefore

$$\frac{P_2}{P_{av}} = \frac{|I_2|^2 R_2}{|E_1|^2/4R_1} = \frac{4R_1R_2\lambda^2}{(R_2+\lambda^2R_1)^2} \tag{4}$$

When $\lambda^2 = R_2/R_1$ the matching is perfect: $P_2/P_{av} = 1$ and the transducer loss is zero. Direct connection of load to generator is simulated by putting $\lambda^2 = 1$, and then $P_2/P_{av} = 4R_1R_2/(R_1+R_2)^2$ while the corresponding current ratio is

$$\left|\frac{I_2}{I_{av}}\right| = \sqrt{\frac{P_2}{P_{av}}} = \frac{\sqrt{(4R_1R_2)}}{R_1+R_2} \tag{5}$$

The loss in dB occasioned by not matching R_2 to R_1 is therefore

$$20\log_{10}\left|\frac{I_{av}}{I_2}\right| = 20\log_{10}\frac{R_1+R_2}{\sqrt{(4R_1R_2)}}, \quad \text{dB} \tag{6}$$

Comment

(1) Insertion loss or gain expresses the effect of a network in a given system that is not necessarily matched; transducer loss or gain expresses it in relation to a system that is ideal in the sense that power transfer would be maximum in the absence of the network, and is thus an absolute measure of the network.

(2) For a passive, reciprocal network the transducer gain cannot exceed unity; for the power leaving the network cannot exceed that entering it, and neither can exceed the maximum, P_{av}. The ideal transformer affords confirmation, for it is a fact of algebra that $4R_2(\lambda^2R_1) \leq (R_2+\lambda^2R_1)^2$.

(3) The expression of general form $\sqrt{(4Z_1Z_2)/(Z_1+Z_2)}$, corresponding to that in equation (6), is known as the *reflection factor* of Z_1 and Z_2. When a two-port network is operated between its image impedances, the reflection loss at each port is zero, and the insertion behaviour is then governed solely by its intrinsic properties as expressed in image parameter terms by its image transfer coefficient, $\frac{1}{2}\log_e(V_2I_2/V_1I_1)$.

ILLUSTRATION 4.12

An ideal transformer is interposed between a generator of emf E_1 and series resistance $R_1 = 10\,\Omega$, and a load of resistance $R_2 = 1\,\Omega$. The transformer has a voltage ratio λ, in the sense from generator to load. Investigate the variation in insertion power ratio, expresssed in decibels, as λ is varied from $\sqrt{10}$ to $\frac{1}{10}\sqrt{10}$.

Interpretation

The conditions with and without the transformer are shown in Figure 4.7.

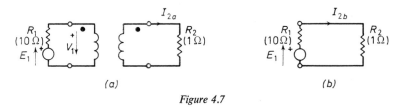

(a)   *(b)*

Figure 4.7

Applying Thevenin's theorem to Figure 4.7(a), noting that the ideal transformer draws no current when open-circuited, so that $V_1 = E_1$ under that condition,

$$I_{2a} = \lambda V_1/(R_2+\lambda^2 R_1) = \lambda E_1/(1+10\lambda^2)$$

Then from Figure 4.7(b), $I_{2b} = E_1/11$, and the insertion power ratio in decibels is therefore

$$20\log_{10}\left|\frac{I_{2a}}{I_{2b}}\right| = 20\log_{10}\frac{11\lambda}{1+10\lambda^2} \tag{1}$$

Evaluating equation (1) for suitable intervals in λ gives Table 4.1 and the graph of Figure 4.8.

TABLE 4.1

λ	λ^2	I_a/I_b	L_{insn}, dB	G_{insn}, dB
$\sqrt{10}$	10	0·344	9·26	—
$\sqrt{10}/2$	5/2	0·669	3·49	—
1	1	1·0	0	0
$2/\sqrt{10}$	2/5	1·39	—	2·86
$1/\sqrt{10}$	1/10	1·74	—	4·81
$1/2\sqrt{10}$	1/40	1·39	—	2·86
1/10	1/100	1·0	0	0
$1/5\sqrt{10}$	1/250	0·669	3·49	—
$1/10\sqrt{10}$	1/1000	0·344	9·26	—

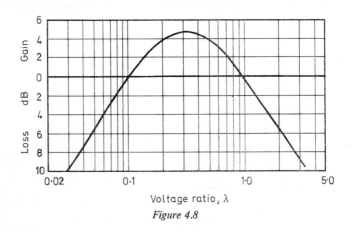

Figure 4.8

Comment

This illustration distinguishes the gain attainable with a passive, reciprocal non-dissipative device from that attainable with an active amplifying device. The transformer does not add energy to the system: its gain is apparent rather than real, in the sense that within a range of ratios (from 1 to $\frac{1}{10}$) it increases the power in the load simply by improving the power-transfer conditions. By contrast, an amplifying device may actually enlarge on the power available from the generator by the conversion of power from an additional source—its dc supply source.

The maximum insertion gain of 4·81 dB occurs when $\lambda^2 = \frac{1}{10}$, the ratio of the resistances. The reactive L-network in Illustrations 4.3 and 4.13 also exhibits gain, of maximum value 4·84 dB at the frequency for which its output impedance is the conjugate of the complex load.

The maximum-power transfer condition thus realised for complex impedances requires a transformation of the form $\lambda^2 Z_1 = Z_2^*$, realisable with an ideal transformer for which λ^2 is real and positive only when arg $Z_1 = -$arg Z_2, or when both arguments are zero. This condition could be satisfied, as an alternative to the L-network in Illustrations 4.3 and 4.13, with a transformer of ratio $\lambda^2 = \frac{1}{10}$ followed by a reactance-annulling capacitor $C_2 = 1/\omega^2 L_2$ in series with the complex load. Such an arrangement would have the advantage of zero phase-shift at ω.

ILLUSTRATION 4.13

Calculate and plot the insertion gain or loss and phase-change of the L-type matching network of Illustration 4.3 over the frequency range $\omega = 0$ to $\omega = 4 \times 10^6$ rad/s.

Interpretation

The calculations are greatly eased by scaling the circuit to a design frequency of $\omega = 1$ rad/s and to a load of series resistance $1\,\Omega$. The scaled values for the corresponding frequency ratio of 10^6 and resistance ratio of 10^2 are

$$C = 3 \times 10^{-9} \times 10^6 \times 10^2 = 0.3 \text{ F},$$

$$L = 2 \times 10^2 \times 10^{-6} \times 10^6 \times 10^{-2} = 2\,\text{H},$$

$$R_1 = 10\,\Omega, \quad R_2 = 1\,\Omega \quad \text{and} \quad L_2 = 1 \text{ H}.$$

It is necessary to calculate the magnitude and argument of the load current with the network present, and with the load and generator directly connected, at several frequencies. Accordingly, a general formula is economical. The two conditions are shown in Figure 4.9.

Writing s for $j\omega$, Figure 4.9(a) is easily solved by applying Thevenin's theorem to the left of the dotted line, giving

$$I_{2a}(s) = \frac{E_1(s)}{1+3s} \bigg/ \left(\frac{10}{1+3s} + 1 + 3s \right)$$

$$= E_1(s)/(9s^2 + 6s + 11)$$

(a)   *(b)*

Figure 4.9

Figure 4.9(b) gives $I_{2b}(s) = E_s(s)/(s+11)$ so that, on replacing s by $j\omega$,

$$\frac{I_{2a}(j\omega)}{I_{2b}(j\omega)} = \frac{11+j\omega}{(11-9\omega^2)+6j\omega}$$

The insertion power ratio is

$$\frac{P_a}{P_b} = \left| \frac{I_{2a}(j\omega)}{I_{2b}(j\omega)} \right|^2 = \frac{121+\omega^2}{(11-9\omega^2)^2+36\omega^2}$$

and the insertion loss or gain in decibels is

$$P_a < P_b, \quad L_{\text{insn}} = 10 \log_{10}(P_b/P_a) \quad \text{dB}$$
$$P_a > P_b, \quad G_{\text{insn}} = 10 \log_{10}(P_a/P_b) \quad \text{dB}$$

The insertion phase-change is

$$\psi = \arg I_{2a}(j\omega)/I_{2b}(j\omega)$$
$$= \tan^{-1}[\omega/11] - \tan^{-1}[6\omega/(11-9\omega^2)]$$

The insertion loss and gain and phase-change are listed in Table 4.2. and plotted as functions of ω in Figure 4.10.

TABLE 4.2

ω	L_{insn}, dB	G_{insn}, dB	ψ, degrees
0	0	0	0
0·5		1·52	−16·33
1·0		4·84	−66·37
1·5	0·828		−124·8
2·0	7·89		−144·1
4·0	21·24		−149·8

Comment

The scaled frequency $\omega = 1$ at which the maximum power transfer condition is satisfied is also the frequency of maximum insertion gain. The network behaves as though it possesses a complex impedance transformation ratio λ^2 (to be consistent with the notation applied to the ideal transformer) such that $\lambda^2 Z_1 = Z_2^*$ when in this case $Z_1 = 10/0$ but $Z_2 = 1+j1$ at the point of maximum power transfer. This condition is, however, accompanied by considerable insertion phase change,

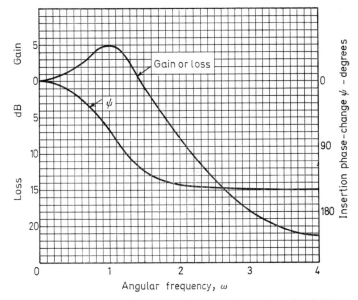

Figure 4.10. Insertion characteristics of L-C matching network of Figure 4.9.

and the frequency characteristics are like those of an asymmetrical low-pass filter whose image impedances become perfectly matched when $\omega = 1$. See also Illustration 4.12.

ILLUSTRATION 4.14

In terms of the general frequency variable s, the filter in Figure 4.11 has the transfer admittance

$$Y_t(s) = I_2(s)/E_1(s) = 1/[2(s^3 + 2s^2 + 2s + 1)]$$

(a) For $s = j\omega$, calculate the insertion loss characteristic for frequencies up to $\omega = 2$ rad/s.

(b) Find the homogeneous equation for $i_2(t)$ and the natural frequencies for the circuit. Show that these are the poles of $Y_t(s)$ and show that the insertion loss can be obtained equivalently from the pole locations on a complex-frequency plane diagram.

Figure 4.11

Interpretation

(a) Let $I_{20}(s)$ denote the current when the 1-Ω generator is connected directly to the 1-Ω load. Then $I_{20}(s) = E_1(s)/2$ and

$$I_{20}(s)/I_2(s) = s^3+2s^2+2s+1 \tag{1}$$

Then, putting $s = j\omega$, the insertion loss is

$$L_{insn} = 10\log_{10}|I_{20}(j\omega)/I_2(j\omega)|^2$$
$$= 10\log_{10}[(1-2\omega^2)^2+\omega^2(\omega^2-2)^2] \tag{2}$$

The values tabulated below are plotted in Figure 4.12.

ω	0	0·5	1·0	1·5	2·0
L_{insn}, dB	0	0·068	3·01	10·9	18·13

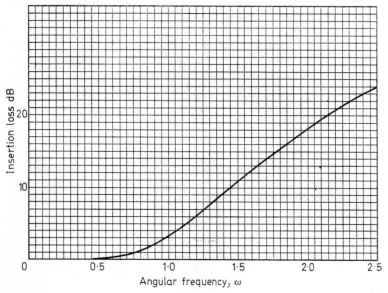

Figure 4.12

(b) Let $i_1(t)$ and $i_2(t)$ denote clockwise mesh currents for an arbitrary excitation $e_1(t)$. Then, using the operator D for compactness,

$$Z_{11}(D)i_1(t)+Z_{12}(D)i_2(t) = e_1(t)$$
$$Z_{21}(D)i_1(t)+Z_{22}(D)i_2(t) = 0 \qquad (3)$$

where

$$Z_{11}(D) = Z_{22}(D) = R+LD+D^{-1}/C = 1+D+D^{-1}/2$$
$$Z_{12}(D) = Z_{21}(D) = -D^{-1}/C = -D^{-1}/2$$

Putting $e_1(t) = 0$ and eliminating $i_1(t)$ gives the homogeneous equation

$$(D^3+2D^2+2D+1)i_2(t) = 0 \qquad (4)$$

which governs the natural behaviour, and the characteristic equation

$$s^3+2s^2+2s+1 = 0 \qquad (5)$$

whose roots are the natural modes or frequencies. Equation (5) may be factorised to give

$$(1+s)(s^2+s+1) = (s-s_1)(s-s_2)(s-s_3) = 0 \qquad (6)$$

where

$$s_1 = -1, \quad s_2 = -\tfrac{1}{2}+j\sqrt{3}/2, \quad s_3 = -\tfrac{1}{2}-j\sqrt{3}/2$$

Comparing $F(s)$ in equation (5) with $F(s)$ in $Y_t(s)$ shows them to be identical, so that replacing s_1, s_2, s_3 by p_1, p_2, p_3,

$$Y_t(s) = \frac{0\cdot5}{(s-p_1)(s-p_2)(s-p_3)}$$

where p_1, p_2 and p_3, corresponding to the natural frequencies, are the poles of $Y_t(s)$, or values of s for which $Y_t(s)$ becomes infinite.

The pole locations for the filter are shown to a scale, originally $5\,\text{cm} = 1\,\text{rad/s}$, in Figure 4.13.

$|F(s)|$ for $s = j\omega$ can be found from the lengths of phasors from a point s on the $j\omega$-axis. For example, when $s = j1$,

$$|F(s)|_{s=j1} = |j1-p_1| \cdot |j1-p_2| \cdot |j1-p_3|$$
$$= (2\cdot6/5)(7\cdot1/5)(9\cdot7/5) = 1\cdot42$$

Then by (1), $L_{\text{insn}} = 20\log_{10}|F(s)| = 3\cdot05$ db, which is in close agreement with the tabulated value.

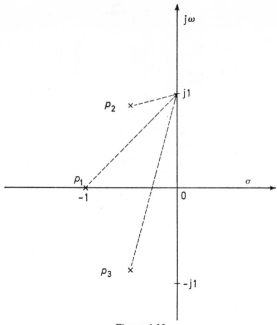

Figure 4.13

Comment

The expression of a driving-point immittance in terms of natural frequencies and the equivalence of poles and zeros to these frequencies have been shown in Illustrations 2.8 and 2.11. In this illustration the role of natural frequencies is extended to embrace the transfer admittance governing the insertion behaviour of a two-port network. This is very important; for it implies the possibility of specifying a desired insertion behaviour as a function of natural frequencies, and finding or *synthesising* a network consistent with that function.

It should be noted that a function having complex natural frequencies, or poles and zeros, cannot pass through infinity or zero at a real frequency; for this is constrained to the imaginary axis and cannot completely cancel a frequency having a real component. This illustration, in which $Y_t(s)$ can never be infinite at a real frequency, might be compared with Illustration 3.6 in which there are zeros on the $j\omega$-axis which can be cancelled at a real frequency.

ILLUSTRATION 4.15

A sinusoidally alternating voltage $V_1(j\omega)$ applied to the input terminals of a two-port network composed of inductors and capacitors, which may be assumed pure, produces a response $I_2(j\omega)$ in a resistor R_2 connected to the output terminals. Show that when R_2 is scaled to $1\ \Omega$, $G_{in}(\omega) = |Y_t(j\omega)|^2$, where $G_{in}(\omega)$ is the frequency-dependent real part ($\mathcal{R}e$) of the complex input admittance $Y_{in}(j\omega)$ of the terminated network, and $Y_t(j\omega)$ is the complex transfer admittance $I_2(j\omega)/V_1(j\omega)$.

Interpretation

Let P_1 and P_2 denote the powers entering and leaving the network. Then,

$$P_1 = |V_1(j\omega)|^2\,\mathcal{R}e\,Y_{in}(j\omega) = |V_1(j\omega)|^2\,G_{in}(\omega)$$

$$P_2 = |I_2(j\omega)|^2\,R_2 = |Y_t(j\omega)V_1(j\omega)|^2\,R_2$$

But as the network itself is non-dissipative, $P_1 = P_2$, and therefore when $R_2 = 1\ \Omega$,

$$|V_1(j\omega)|^2 G_{in}(\omega) = |Y_t(j\omega)V_1(j\omega)|^2$$

or

$$G_{in}(\omega) = |Y_t(j\omega)|^2$$

Comment

The reader may derive easily an equivalent relationship in terms of an excitation current $I_1(j\omega)$ and a response $V_2(j\omega)$ across R_2. Then, for $R_2 = 1\ \Omega$, $\mathcal{R}e\,Z_{in}(j\omega) = R_{in}(\omega) = |Z_t(j\omega)|^2$ where $Z_t(j\omega) = V_2(j\omega)/I_1(j\omega)$.

Most brilliant ideas are founded on simple basic principles. This is true of the synthesis technique for two-port reactance networks having prescribed transfer functions published by S. Darlington in 1939, which is thought of as the fore-runner of modern synthesis. The method rests on the simple correlation between $|Z_t(j\omega)|^2$ and $\mathcal{R}e\,Z_{in}(j\omega)$. When the desired transfer function (filter characteristic) is specified in the form $|Z_t(j\omega)|^2$, $\mathcal{R}e\,Z_{in}(j\omega)$ is prescribed; but $\mathcal{R}e\,Z_{in}(j\omega)$ is uniquely related to the complex $Z_{in}(j\omega)$, which may be found from $\mathcal{R}e\,Z_{in}(j\omega)$ by mathematical procedures originating in the theory of complex variables (this is the relatively difficult part). $Z_{in}(j\omega)$ thus found may be expressed as a ratio of polynomials in terms of the general frequency variable s, and the terminated network can then be synthesised by relatively simple techniques available for the realisation of driving-point (or input) impedance functions.

ILLUSTRATION 4.16

The input or driving-point admittance of a two-port pure reactance network terminated with a resistor of $1\ \Omega$ is

$$Y_{in}(s) = \frac{3s^3 + 6s^2 + 6s + 3}{2s^2 + 4s + 3} \tag{1}$$

What is the magnitude of the transfer admittance when $s = j1$?

Interpretation

When $s = j\omega$, corresponding to the sinusoidal steady state,

$$Y_{in}(s) = Y_{in}(j\omega); \quad \text{and for} \quad s = j1,$$

$$Y_{in}(j1) = \frac{-j3 - 6 + j6 + 3}{-2 + j4 + 3} = \frac{-3 + j3}{1 + j4}$$

Rationalising,

$$Y_{in}(j1) = \frac{(-3 + j3)(1 - j4)}{(1 + j4)(1 - j4)} = \frac{9 + j15}{17}$$

whence

$$\mathcal{R}e\ Y_{in}(j1) = \tfrac{9}{17}$$

and therefore, since $|Y_t(j\omega)|^2 = \mathcal{R}e\ Y_{in}(j\omega)$,

$$|Y_t(j1)| = 3/\sqrt{17}$$

Comment

The given driving-point admittance corresponds to a transfer admittance magnitude specification of the form

$$|Y_t(j\omega)| = \left| \frac{I_2(j\omega)}{V_1(j\omega)} \right| = \frac{1}{\sqrt{(1 + \omega^6)}} \tag{2}$$

This is a *Butterworth-type* response, for which the general form is

$$|T(j\omega)| = \frac{H}{\sqrt{(1 + \omega^{2n})}_{n = 1, 2, \ldots}} \tag{3}$$

The Butterworth (also known as *maximally flat*) form of response is one of the common specifications for low-pass filter synthesis.

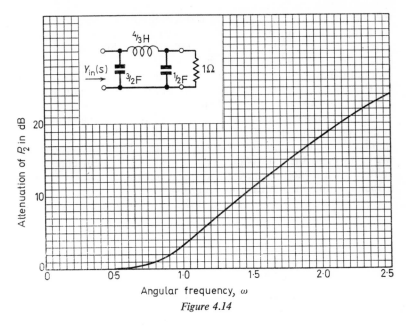

Figure 4.14

The network conforming to the specifications of equations (2) and (1) is shown in Figure 4.14 together with the power P_2 in R_2 relative to that at $\omega = 0$, plotted as a function of ω on a decibel scale. The curve follows the equation

$$10 \log_{10}(P_{2(\omega)}/P_{2(0,)}) = 10 \log_{10}[1/(1+\omega^6)]$$

ILLUSTRATION 4.17

What is meant by the term *scattering coefficient* as applied to a multi-port network? Derive expressions for the current and voltage scattering coefficients for the case of a generator of impedance Z connected to one port of a network at which the input impedance is Z_{in}.

Interpretation

The term *scattering coefficient* is now preferred to *reflection co-efficient*, which has been applied for a long time not only to distributed systems such as transmission lines and waveguides, but also to lumped networks. In a distributed system, space and time are integrated into a wave-mechanism of transmission, and reflection is physically realistic;

12

for the incidence of a wave with a discontinuity in the propagation medium initiates an observable reverse or reflected wave. A lumped network does not, however, exhibit reflection in this physical sense; but a useful analogy exists, in the sense that the mismatching of a generator may be likened to an immitance discontinuity responsible for the 'reflection' of power that would otherwise have contributed to the maximum acceptable by the load.

The scattering coefficient for a distributed system is the ratio of the amplitude of the reflected wave to that of the incident, at the discontinuity. A similar definition is applicable to a discrete electrical network, through the contrivance of reflected and incident currents or voltages; but as no waves exist and the reflection concept is then artificial, a simple theoretical approach is valid.

Figure 4.15 depicts in dual forms the junction of a generator with one port of a network.

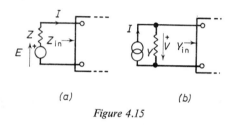

(a)   (b)

Figure 4.15

Let the responses I and V in Figure 4.15(a) and (b) be resolved into hypothetical incident and reflected components I_i, I_r and V_i, V_r, so that for (a) $I = I_i + I_r$ and for (b) $V = V_i + V_r$. The incident components are the current and voltage that would have existed at the network port had its immitance been the image of the source immitance, and are

$$I_i = E/2Z \quad \text{and} \quad V_i = I/2Y$$

in the positive senses of the excitations. But $Z_{in} \neq Z$, and $Y_{in} \neq Y$, so that the actual responses in the positive senses of the excitations are

$$I = E/(Z + Z_{in}) \quad \text{and} \quad V = I/(Y + Y_{in})$$

The depature from the incident values is attributable to the hypothetical *reflected* or *scattered components*, or since $I = I_i + I_r$ and $V = V_i + V_r$

$$I_r = -(I_i - I) = E(Z - Z_{in})/[2Z(Z + Z_{in})]$$

and

$$V_r = -(V_i - V) = I(Y - Y_{in})/[2Y(Y + Y_{in})]$$

The *scattering coefficients* are therefore

$$S_I = \frac{I_r}{I_i} = \frac{Z - Z_{in}}{Z + Z_{in}} \tag{1}$$

and

$$S_V = \frac{V_r}{V_i} = \frac{Y - Y_{in}}{Y + Y_{in}} \tag{2}$$

Substituting $Z = 1/Y$ in equation (2) gives

$$S_V = -\frac{Z - Z_{in}}{Z + Z_{in}} = -S_I \tag{3}$$

Comment

(1) The formulae derived are the same as the reflection coefficients for a mis-matched transmission line, though in that context a more classical wave-propagation approach is desirable for rigour.

(2) The expressions have the form difference over sum; but the signs are not readily remembered. However, limiting conditions afford an easy check: as $Z_{in} \to \infty$ $I \to 0$, and this requires $I_r \to -I_i$. The inference is therefore S_I negative for $Z_{in} > Z$. Similarly, as $Y_{in} \to \infty$ or $Z_{in} \to 0$, $V \to 0$, and the inference is S_V negative for $Z_{in} < Z$.

(3) The usefulness of the scattering coefficients for a discrete network is mathematical, for they absorb both generator and input immittances into a single symbol through which external relations, especially power, can be neatly and clearly interpreted. See Illustration 4.18.

(4) Scattering coefficients may be associated with each port of a multi-port network. For example, a two-port network operated between impedances Z_1 and Z_2 may be attributed incident and reflected voltage components V_{i1}, V_{r1}, and V_{i2}, V_{r2} at the input and output ports. These may be interrelated by four scattering coefficients in the form of a *scattering matrix*, for which the matrix equation has the form

$$\begin{bmatrix} S_{11} & S_{12} \\ S_{21} & S_{22} \end{bmatrix} \cdot \begin{bmatrix} V_{i1} \\ V_{i2} \end{bmatrix} = \begin{bmatrix} V_{r1} \\ V_{r2} \end{bmatrix}$$

where each coefficient is defined under a short-circuit condition, e.g.

$$S_{11} = \frac{V_{r1}}{V_{i1}} \bigg|_{V_{i2} = 0} , \quad S_{12} \frac{V_{r1}}{V_{i2}} \bigg|_{V_{i1} = 0} \quad \text{etc.}$$

ILLUSTRATION 4.18

A two-port network of pure inductors and capacitors is connected between a generator of emf E_1 and series resistance R_1, and a load of resistance R_2. Show that the transducer power-gain of the network can be expressed in the form

$$G(\omega) = 1 - |S(j\omega)|^2$$

where S is the current or voltage scattering coefficient at the input port.

Interpretation

As the network is non-dissipative, the load-power P_2 equals the input P_1. This may be resolved into hypothetical incident and reflected components P_i and P_r, where P_i is the available power $P_{av} = E_1^2/4R_1$ (see Illustration 4.11), and P_r may be expressed as $P_r = |V_r I_r| \cos \phi$, where V_r and I_r are the reflected voltage and current and ϕ is their phase displacement.

Let $S(j\omega) = |S(j\omega)| \underline{/\theta}$ denote the voltage scattering coefficient. Then for current it is $-S(j\omega) = -|S(j\omega)| \underline{//\theta}$, which may be written as $|S(j\omega)| \underline{/\theta + \pi}$. The phase displacement ϕ is therefore $\phi = (\theta + \pi) - \theta = \pi$, and

$$P_r = |V_r I_r| \cos \phi = |V_i S(j\omega)| \cdot |I_i S(j\omega)| \cos \pi$$
$$= -|V_i I_i| \cdot |S(j\omega)|^2 = -P_i |S(j\omega)|^2$$

Then,

$$P_2 = P_1 = P_i + P_r = P_i(1 - |S(j\omega)|^2)$$

But P_i equals the available power P_{av}, and the transducer gain is therefore

$$G(\omega) = \frac{P_2}{P_{av}} = 1 - |S(j\omega)|^2$$

Comment

In Illustration 4.15 it is shown that the transfer admittance $Y_t(j\omega) = I_2(j\omega)/V_1(j\omega)$ for a reactance network scaled to a $1\,\Omega$ terminating resistor is related to the input admittance in the form $|Y_t(j\omega)|^2 = G_{in}(\omega)$. It has been shown how use of the scattering coefficient concept extends this very simply to include the generator resistance, in the form of an elegant formula for the transducer gain between finite resistances. This is the essence of Darlington's procedure for the synthesis of double-terminated two-port reactance networks.

It is also noteworthy that as the scattering coefficient cannot exceed unity, the expression $1 - |S(j\omega)|^2$ affords further evidence that the transducer gain of a passive reciprocal network cannot exceed unity. (See Illustration 4.11).

CHAPTER 5

Examples of non-linearity and the response of networks to non-sinusoidal waveforms

INTRODUCTION

Non-linearity means failure of a network or system to respond in constant proportion to its excitation. A resistor R for which the time-domain response is $i(t) = v(t)/R$ is linear when i is in fixed proportion to v at all instants; a capacitor C, for which $v(t) = q(t)/C$, is linear when v is in constant proportion to the charge q; and its dual, a pure inductor L for which $i(t) = \Lambda(t)/L$, is linear when i is in constant proportion to the flux-linkage Λ. Non-linearity thus implies that the circuit elements themselves are not constant, as assumed in linear network analysis, but are additional variables in the Kirchhoff-law equations. This makes non-linear analysis difficult, except for simple cases involving only one non-linear element, or at the most a few such elements, approachable in a piecewise fashion.

Non-linearity has attributes that are at variance with one another: it is undesirable, as in the performance of transducers and amplifiers; it is useful, as in the processes of modulation or frequency-changing and for the generation of frequency spectra; and, perhaps above all, it is fundamental to restraint in growth, as in the stabilisation of amplitude in an oscillator. Without non-linearity as a fundamental natural phenomenon, much that is possible would be explosive and impossible. A good general treatment of non-linear distortion is given in chapter 5 of the book 'Communication Systems Analysis' (see Bibliography).

Practical resistors, capacitors and inductors are substantially linear within their working ranges, and are usually treated as linear in analysis. Inductors and transformers with magnetic cores, however, are less linear, and may be very non-linear at high flux densities. There are forms of resistor and capacitor, such as the thermistor and varactor diode, that are designed to be non-linear, for specific applications. The thermistor is often used for automatic amplitude control, and the varactor diode has many applications, including automatic tuning and frequency modulation.

Non-linearity in electronic circuits resides mainly in the electronic devices themselves, which are all non-linear. A transistor exhibits appreciable non-linearity in all its driving-point and transfer parameters (see Illustration 5.5), while a tunnel diode exhibits a range of negative resistance (Illustration 5.6). Even the most perfect rectifier must be regarded as non-linear, in the sense that it is unilateral and conducts for an excitation in one direction only (see Illustration 5.7, 5.8 and 5.9).

ILLUSTRATION 5.1

An amplifier has an input-output transfer characteristic expressed by

$$i = I_Q + a_1 v + a_2 v^2$$

where i is the current output into a load resistance, v is the input signal voltage, and I_Q is the dc operating current of the amplifier under quiescent ($v = 0$) conditions.

If $v = \hat{V} \sin \omega t$, show (a), that the ratio of the second harmonic distortion product $\hat{I}_{2\omega}$ to the component $\hat{I}_\omega$ at the fundamental frequency ω is $(a_2/2a_1)\hat{V}$; and (b), that

$$\frac{\hat{I}_{2\omega}}{\hat{I}_\omega} = \frac{i_{max} + i_{min} - 2I_Q}{2(i_{max} - i_{min})}$$

where i_{max} and i_{min} are the maximum and minimum values between which the output current i varies during a cycle of the input voltage v.

Interpretation

(a) Substituting for v,

$$i = I_Q + a_1 \hat{V} \sin \omega t + a_2 (\hat{V} \sin \omega t)^2$$
$$= I_Q + a_1 \hat{V} \sin \omega t + a_2 \hat{V}^2 (1 - \cos 2\omega t)/2$$
$$= \left(I_Q + \frac{a_2 \hat{V}^2}{2}\right) + a_1 \hat{V} \sin \omega t - \frac{a_2 \hat{V}^2 \cos 2\omega t}{2}$$

whence

$$\frac{\hat{I}_{2\omega}}{\hat{I}_\omega} = \frac{a_2\hat{V}^2}{2}\cdot\frac{1}{a_1\hat{V}} = \frac{a_2}{2a_1}\hat{V}$$

(b) The current is instantaneously a maximum when $\omega t = \pi/2$ ($\sin\omega t = 1$, $\cos 2\omega t = -1$) and a minimum when $\omega t = 3\pi/2$ ($\sin\omega t = -1$, $\cos 2\omega t = -1$). Then,

$$i = i_{max} = I_Q + a_1\hat{V} + a_2\hat{V}^2$$

and

$$i = i_{min} = I_Q - a_1\hat{V} + a_2\hat{V}^2$$

whence

$$a_1\hat{V} = (i_{max} - i_{min})/2$$

$$a_2\hat{V}^2 = (i_{max} + i_{min} - 2I_Q)/2$$

and thus

$$\frac{\hat{I}_{2\omega}}{\hat{I}_\omega} = \frac{a_2\hat{V}}{2a_1} = \frac{i_{max} + i_{min} - 2I_Q}{2(i_{max} - i_{min})} \qquad (1)$$

Comment

(a) Note the dependence of the distortion ratio on the input voltage. Provided $a_2 \ll a_1$, the departure from linearity in the characteristic is not great and the distortion may be negligible for small values of v. This is the implication in assuming linear behaviour of an amplifying device when applying linear network analysis under small signal conditions.

The non-linear transfer characteristic implies a voltage-dependent transfer conductance. This is readily seen by differentiating the power series, to give

$$\frac{\partial i}{\partial v} = a_1 + 2a_2v = G_1 + G_2(v)$$

(b) The algebraic pattern of the power series is quite general, and a distortion formula such as equation (1), derived from the particular, conventional form $i = f(v)$, is adaptable to any analogous transfer or driving-point function.

ILLUSTRATION 5.2

A two-terminal non-linear circuit element is connected in series with a linear resistor R of resistance 100 Ω. When a voltage v is applied to the series combination, the current is expressed by

$$i = 10^{-3}v + 10^{-4}v^2, \quad \text{A}$$

In a particular application, $v = 5 \sin \omega t + 5 \sin 10\omega t$. Find, (a) the mean voltage across R, (b) the power dissipated in R, (c) the amplitudes and angular frequencies (in terms of ω) of the alternating components of voltage across R.

Interpretation

The instantaneous voltage across R is

$$v_R = Ri = 100(10^{-3}v + 10^{-4}v^2)$$
$$= 0.5(\sin \omega t + \sin 10\omega t)$$
$$+ 10^{-2}(25 \sin^2 \omega t + 50 \sin \omega t \sin 10\omega t + 25 \sin^2 10\omega t)$$
$$= 0.5(\sin \omega t + \sin 10\omega t)$$
$$+ 0.125[1 - \cos 2\omega t + 2(\cos 9\omega t - \cos 11\omega t) + 1 - \cos 20\omega t]$$

(a) The mean voltage is the dc component,

$$V_0 = 0.125 + 0.125 = 0.25 \text{ V}$$

(b) The coefficients of the components represent their peak values, and the rms value is thus given by

$$V = \sqrt{\left\{ 0.25^2 + \frac{2(0.5^2 + 0.125^2 + 0.25^2)}{2} \right\}} = 0.625 \text{ V}$$

The power dissipated in R is therefore

$$P = V^2/R = 3.91 \text{ mW}$$

(c) The angular frequencies and their amplitudes are

Frequency component	Amplitude, V
ω	0.5
10ω	0.5
2ω	0.125
20ω	0.125
9ω	0.25
11ω	0.25

Comment

The non-linearity is responsible for the production of four new frequency components. These comprise harmonics (2ω, 20ω) of the original frequencies and sum (11ω) and difference (9ω) components. Such new components are distortion products, minimised in normal amplifier application, but deliberately invoked in processes of frequency changing such as modulation, detection, and the generation of carrier frequencies for multi-channel communication systems.

ILLUSTRATION 5.3

A non-linear passive device N is connected in series with a resistor R to a supply of voltage $e = \hat{E} \sin \omega t$ and negligible internal impedance. The current is

$$i = \hat{I}_1 \sin (\omega t + \phi_1) + \hat{I}_3 \sin (3\omega t + \phi_3)$$

Determine

(a) the voltage v across the non-linear device N as a function of time t, in terms of the given quantities;

(b) the third-harmonic component of the voltage v and its phase relation to the corresponding component voltage across R.

Hence, or otherwise, determine the mean power supplied to N. Comment on the significance of the results.

(L.U., Part 2, Electrical Theory and Measurements).

Interpretation

(a) To satisfy Kirchhoff's voltage law, $\Sigma v = 0$ and therefore

$$Ri + v - \hat{E} \sin \omega t = 0$$

whence

$$v = \hat{E} \sin \omega t - Ri$$
$$= \hat{E} \sin \omega t - R\hat{I}_1 \sin (\omega t + \phi_1) - R\hat{I}_3 \sin (3\omega t + \phi_3)$$

(b) The third harmonic component in v_N is equal in magnitude but opposite in phase to the corresponding voltage across R. This must be so in order to give zero sum with the supply, which contains no third harmonic component.

The mean power supplied to N is

$$P_N = P_{\text{in}} - P_R$$

where P_{in}, the net power from the source, is compounded from the fundamental frequency components alone; and P_R, the power dissipated in the resistor, is determined from the rms value of the complex current. Thus,

$$P_N = \frac{\hat{E}}{\sqrt{2}} \cdot \frac{\hat{I}_1}{\sqrt{2}} \cos \phi_1 - R\left[\left(\frac{\hat{I}_1^2 + \hat{I}_3^2}{2}\right)^{1/2}\right]^2$$

$$= \tfrac{1}{2}\hat{E}\hat{I}_1 \cos \phi_1 - \tfrac{1}{2}(\hat{I}_1^2 + \hat{I}_3^2)R$$

Comment

(a) Kirchhoff's voltage and current laws, $\Sigma v = 0$ for a loop and $\Sigma i = 0$ for a node, must be obeyed even when non-linear elements are present. This is the consideration that dictates the mutual cancellation of the third harmonic components of voltage across the resistor and the non-linear element.

(b) A mean power can exist only as the result of the interaction of voltages and currents of identical frequency and (thus) constant phase difference. In this problem the third harmonic component of current is therefore of no significance in calculating the net input power.

ILLUSTRATION 5.4

The voltage-current characteristic of a non-linear circuit element N is expressed by

$$i = 0{\cdot}1v + 0{\cdot}02v^2, \text{ A}$$

(a) The element N is connected in series with a resistor $R = 3\,\Omega$ across a constant-voltage dc supply $V_s = 6$ V. Find the current i, and the voltage v across N.

(b) A sinusoidally alternating emf $E = 2$ V rms is connected in series with the dc supply. What then are the instantaneous maximum and minimum values of voltage across N and current in the circuit?

Interpretation

Under dc conditions only, the voltage across N is $v = V_s - iR$; but with the alternating emf superimposed, it is $v = v_s - iR$, where $v_s = V_s + \hat{E} \sin \omega t$. Substituting for v in the $v-i$ equation for N yields a quadratic equation that may be solved for i. But this approach is tedious, especially with the superimposed alternating emf; and in the case of a non-linear element specified by a higher-order power series, the work may become almost intractable.

A simpler alternative approach is to obtain the simultaneous solutions for v and i graphically. This is done merely by plotting the equations

$$i = (v_s - v)/R$$

and

$$i = 0.1v + 0.02v^2$$

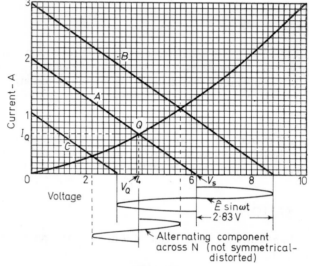

Figure 5.1

as functions of v: the solutions for i and v are then the coordinates of the point of intersection. This approach is moreover general, for it does not require the equation for the non-linear element if a curve for its $v-i$ characteristic is available from experimental data, and the curve for the resistor is simply a straight line intersecting the current axis at $i = v_s/R$ and the voltage axis at $v = v_s$, where v_s is the supply voltage at a particular instant in time.

(a) The curves are shown in Figure 5.1. Under dc conditions, the resistance line A intersects the current axis at $i = V_s/R = 2$ A, and the voltage axis at $v = V_s = 6$ V. The coordinates of the point of intersection give $i = I_Q = 0.7$ A and $v = V_Q = 3.9$ V.

(b) The sinusoidal emf $E = 2$ V rms has a peak value $\hat{E} = 2\sqrt{2} = 2.83$ V. When it is connected in series with the dc supply V_s, the instantaneous supply voltage is

$$v_s = V_s + \hat{E}\sin \omega t = 6 + 2.83 \sin \omega t$$

and when $\sin \omega t = +1$ and -1, the corresponding maximum and minimum values are

$$v_{s\,max} = 8 \cdot 83 \text{ V}, \quad v_{s\,min} = 3 \cdot 17 \text{ V}.$$

B and C are the resistance lines corresponding to these voltages, and from their intersections with the curve for N,

$$v_{max} = 5 \cdot 4 \text{ V}, \quad i_{max} = 1 \cdot 15 \text{ A}$$
$$v_{min} = 2 \cdot 2 \text{ V}, \quad i_{min} = 0 \cdot 32 \text{ A}$$

By applying equation (1), Illustration 5.1, $I_{2\omega} \simeq 4\% \, I_\omega$.

Comment

The graphical approach illustrated is very important in electronics. No semiconductor device or electron tube has a truly linear volt-ampere characteristic, although for small increments in voltage and current about an operating point, it is often treated as linear, since the departure from linearity may be imperceptible over a small range. In the case of a device intended to deliver a substantial output power, the variations in voltage and current are large, and the effects of non-linearity are then important.

In this context the linear resistance component is usually the load in which the power is dissipated, and the straight-line graph corresponding to $i = (v_s - v)/R$ is called a *load line* (see Illustration 5.5).

An important related technique is the *Phase-Plane* and *Phase-Trajectory*. This is a graphical method of determining the transient response of a non-linear system or circuit containing energy-storage elements (L and C in the electrical case, mass and compliance in the analogous mechanical case).

ILLUSTRATION 5.5

(a) What are the principal non-linearities of a transistor, and in particular, what is the significance of non-linearity in its base-emitter input characteristic in respect of an applied voltage of given waveform?

(b) A common-emitter transistor amplifier has a resistance load that is coupled into the collector circuit with a transformer having negligible winding resistances and substantially ideal transformation properties in the working frequency range. Under these conditions, what governs the quiescent operating point and the response in the load on the amplifier to a sinusoidal current excitation traversing base and emitter?

180

Interpretation

(a) All the characteristics are non-linear to varying degrees. But an $I_C - V_{CE}$ characteristic for I_B constant is substantially linear beyond a region bounded by $V_{CE} \simeq 1$ V (see Figure 5.2). The transfer relation $I_C = h_{fe}I_B$ is moderately linear, for h_{fe} is not severely dependent either on I_C or on V_{CE}. Silicon transistor type BC 107, for example, shows an overall variation in h_{fe} of order 15% as I_C varies from 1 to 10 mA at $V_{CE} = 5$ V, while small-signal values vary only about 5% for this current range. Doubling V_{CE} (a large change by small-signal standards) changes h_{fe} by less than 10%.

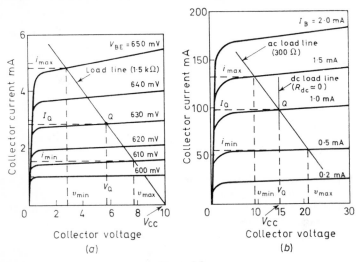

Figure 5.2

Of particular significance is the input resistance non-linearity. For type BC 107 h_{ie} at $V_{CE} = 5$ V varies from about 7 kΩ at $I_C = 1$ mA to 1·2 kΩ at $I_C = 10$ mA. The base current response $i_b(t)$ to an excitation $v_{be}(t)$ therefore suffers distortion similar to that in Figure 5.1, Illustration 5.4. The non-linearity in $i_c(t)$ and $v_{ce}(t)$ as functions of $v_{be}(t)$ is implied in Figure 5.2(a) by the divergence of the $I_C - V_{CE}$ curves for equal increments in V_{BE}. The load line shown is for a collector load resistor of 1·5 kΩ and a collector supply V_{CC} of 10 V. With V_{BE} biassed at 630 mV, the intersection of the load line with the 630 mV curve defines the quiescent operating point Q, at which the collector current and voltage have the steady values $I_Q = 2·8$ mA and $V_Q = 5·8$ V. A superimposed sinusoidal excitation of peak value 20 mV then moves the intersection to $i_{max} = 4·8$ mA, $v_{min} = 2·8$ V and to $i_{\min} = 1{\cdot}5$ mA, $v_{\max} = 7{\cdot}8$ V. By equation (1), Illustration 5.1, the second harmonic distortion is about 11%.

(b) Figure 5.2(b) shows $I_C - V_{CE}$ curves for transistor type BC 187 in the customary form as functions of base current I_B. A more linear response to $i_b(t)$ is predictable from the more uniform spacing of the curves.

A transformer-coupled resistance load R_L causes no dc voltage-fall between supply and collector, so that if the transformer winding resistance is negligible the steady collector voltage equals the supply voltage V_{CC}. The quiescent point Q is then that at which a vertical line from V_{CC} (the dc load line for zero resistance) intersects the $I_C - V_{CE}$ curve for the chosen base-bias current. To an ac excitation, however, the transformer presents in the collector circuit a resistive impedance $(N_p/N_s)^2 R_L$, where N_p/N_s is the turns-ratio. A distinction must therefore be made between an ac load line representing this impedance, and a dc load line governed solely by the dc resistance of the transformer primary winding, which may often be negligible.

The quiescent operating point Q in Figure 5.2(b) for a base-bias current of 1 mA is at $V_Q = V_{CC} = 15$ V (ignoring dc resistance) and $I_Q = 98$ mA, and the ac load line is for a resistive impedance of $300\,\Omega$. A sinusoidal base-current excitation of peak value $0{\cdot}5$ mA gives $i_{\max} = 133$ mA, $v_{\min} = 9$ V and $i_{\min} = 55$ mA, $v_{\max} = 21$ V. The output power, into the transformer and thence to the load, is $\Delta v/2\sqrt{2} \times \Delta i/2\sqrt{2}$ or $(v_{\max} - v_{\min})(i_{\max} - i_{\min})/8$, which gives $117\,\mathrm{mW}$. The distortion, too small for a reliable graphical estimate, is probably less than 2%.

Comment

A junction transistor is often classed as a current-operated device; for while it loads an input voltage source appreciably and in a non-linear way that is important except for small voltages, its current transfer properties are reasonably linear. These factors influence transistor voltage and power amplifier design, and it is common to minimise all non-linearities with negative feedback. For example, input resistance is raised by the feedback due to an emitter resistor (see Illustration 1.18); and most integrated circuits are used as operational amplifiers, in which the effective voltage gain is set by the ratio of two linear resistors and is almost independent of the transistors (see Illustration 1.10). Transistor non-linearities are complicated further by the temperature dependence of the parameters.

The distinction between ac and dc load lines applies to any load other than a physical resistor. A parallel-tuned circuit, for example, has negligible dc resistance but a large ac *(dynamic)* resistance $(Q\omega L)$ at its parallel-resonant frequency.

182

ILLUSTRATION 5.6

In Figure 5.3 is shown the current-voltage characteristic of a non-linear circuit element exhibiting a region of negative resistance. As indicated, this element N is connected in series with a linear resistor R across a variable unidirectional emf E having a negligible internal resistance.

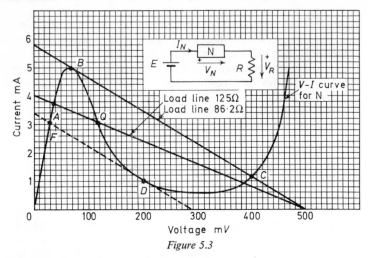

Figure 5.3

(a) If $R = 125\,\Omega$ and E is raised from 0 to 500 mV, estimate graphically V_N, I_N and V_R when $E = 500$ mV.

(b) If E is maintained at 500 mV, estimate the value to which R must be altered in order that the circuit may just *trigger* (change abruptly) to a new state of equilibrium, and estimate V_N, I_N and V_R for this new state.

(c) Let R be fixed at its new value, with the circuit in its new state of equilibrium. If E is then reduced from 500 mV, at what value of E will the circuit trigger again, and between what values will V_N, I_N and V_R change?

Interpretation

(a) As E is raised from zero, the load line, of constant slope corresponding to $R = 125\ \Omega$, moves across the characteristic of N; and when $E = 500$ mV, its intercepts are at $I = 500/125 = 4$ mA on the current axis, and at 500 mV on the voltage axis. V_N, I_N and V_R are given by the coordinates of the intersection at A, and they are approximately

$$V_N = 37\,\text{mV}, \quad I_N = 3{\cdot}7\,\text{mA}, \quad V_R = E - V_N = 463\,\text{mV}$$

(b) If R is decreased, the intersection moves from A towards B, at which point it is tangential. An infinitesimal further reduction in R transfers the contact at B to an intersection at C, with corresponding abrupt changes or *triggering* in voltage and current values to a new state of equilibrium. The critical resistance for triggering is thus given by the slope of the tangent at B, and is $R = 86 \cdot 2 \, \Omega$. The voltages and current for the new state at the point C are approximately

$$V_N = 400 \text{ mV}, \quad I_N = 1 \cdot 15 \text{ mA}, \quad V_R = 100 \text{ mV}$$

(c) When $R = 86 \cdot 2 \, \Omega$ and the circuit is triggered to C, reducing E moves the $86 \cdot 2 \, \Omega$ load line across the characteristic towards the point of tangency D. At this point the approximate values are

$$E = 286 \text{ mV}, \quad V_N = 200 \text{ mV}, \quad I_N = 1 \cdot 0 \text{ mA}, \quad V_R = 86 \text{ mV}$$

An infinitesimal further reduction in E triggers the circuit to F at which, approximately,

$$V_N = 30 \text{ mV}, \quad I_N = 3 \cdot 0 \text{ mA}, \quad V_R = 256 \text{ mV}$$

Comment

(1) A point of intersection such as Q in Figure 5.3 can be reasoned as unstable, by considering a small increment in voltage or current about 0.* The stable point is therefore always one such as A or C.

(2) A resistance characteristic of the kind shown in Figure 5.3, embracing a region of negative slope, is an intrinsic feature of electronic trigger or bistable circuits. A basic example is the Eccles–Jordan circuit, which exhibits a negative resistance characteristic of symmetrical form between anodes in the 2-valve circuit, and between collectors in the 2-transistor equivalent.

Such circuits are also the basis of relaxation oscillators (such as the free-running multivibrator), in which the triggering is a cyclic function of the charging and discharging of capacitors through resistors. Negative resistance characteristics may also be used to realise sinusoidal oscillations, by neutralising the losses in tuned circuits. Figure 5.3 is, in fact, the characteristic of a semiconductor tunnel diode, which is applicable to the generation of oscillations at very high frequencies.

* B.D.H. Tellegen has shown that a complex-frequency plane approach is necessary for rigour (Stability of negative resistances, B.D.H. Tellegen, London 1971 International Symposium on Electrical Network Theory)

ILLUSTRATION 5.7

Figure 5.4(a) is the basic arrangement of a switching circuit in which the conduction of the rectifiers is controlled by the low-resistance high-amplitude source $\hat{E}_0 \sin \omega_0 t$.

Assuming that the rectifiers have negligible forward (conducting) resistances and very high reverse resistances, show that when $\omega < \omega_0$, the waveform of the low-amplitude source $\hat{E} \sin \omega t$ is sampled by pulses of duration π/ω_0 as indicated in Figure 5.4(b); and that the resultant current $i_s(t)$ in R_2 has a frequency spectrum of the form ω, $\omega_0 \pm \omega$, $3\omega_0 \pm \omega$, $5\omega_0 \pm \omega$, ... Determine the amplitudes of the frequency components when $R_1 = R_2 = 1 \text{ k}\Omega$ and $\hat{E} = 1 \text{ V}$.

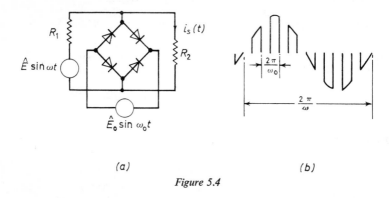

(a)   (b)

Figure 5.4

Interpretation

Let $i(t)$ denote $\hat{I} \sin \omega t = \hat{E} \sin \omega t/(R_1 + R_2)$, the current that would flow continuously if the rectifiers were absent. The rectifiers are connected, however, so that they are non-conducting during one half-period of $\hat{E}_0 \sin \omega_0 t$ and conducting during the other. Their action, when assumed perfect, is therefore like that of a switch that short-circuits R_2 every alternate half-period of $\hat{E}_0 \sin \omega_0 t$, so that $i(t)$ flows in alternate intervals of duration π/ω_0 as indicated in Figure 5.4(b).

The on–off switching action is represented by $i(t) \times 1$ for on, and $i(t) \times 0$ for off. The periodic action is accordingly represented by a current $i_s(t) = i(t) \times f(t)$, where $f(t)$ is a periodic function of time, of period $2\pi/\omega_0$, having unit amplitude for $0 < t < \pi/\omega_0$ and zero amplitude for $\pi/\omega_0 > t > 2\pi/\omega_0$. The Fourier series for this rectangular-wave function is

$$f(t) = \frac{1}{2} + \frac{2}{\pi}\left(\sin \omega_0 t + \frac{1}{3}\sin 3\omega_0 t + \frac{1}{5}\sin 5\omega_0 t + \dots\right)$$

Products of the form $2 \sin n\omega_0 t \cdot \sin \omega t$ give terms of the form $\cos(n\omega_0 - \omega)t - \cos(n\omega_0 + \omega)t$ so that

$$i_s(t) = i(t) \cdot f(t) = \hat{I} \sin \omega t \cdot f(t)$$

resolves into a spectrum of components having frequencies ω, $\omega_0 \pm \omega$, $3\omega_0 \pm \omega$, $5\omega_0 \pm \omega$, ..., and corresponding amplitudes $\hat{I}/2$, $\hat{I}/\pi$, $\hat{I}/3\pi$, $\hat{I}/5\pi$, ... For the given figures, the amplitudes in mA are therefore $\frac{1}{4}$, $\frac{1}{2}\pi$, $\frac{1}{6}\pi$, $\frac{1}{10}\pi$, ...

Comment

Figure 5.4(a) is much used as a modulator or frequency-changer. It is one of a variety of rectifier circuits available for this purpose, and illustrates again the inception of new frequencies in a system that is not linear. While this case may look quite different from Illustration 5.2, it is nevertheless a manifestation of non-linearity in a different form: $i_s(t)$ is non-linearly related to $\hat{E} \sin \omega t$ by $f(t)$, which is a non-linear function of $\hat{E}_0 \sin \omega_0 t$.

The desired side-frequency in practice is usually $\omega_0 \pm \omega$. As either of these frequencies is so far removed from ω and $3\omega \pm \omega$, it is very easy to filter out.

ILLUSTRATION 5.8

A high-frequency amplifying device operated under non-linear conditions may be represented as a current source, of negligible shunt conductance, having the form

$$i(t) = \frac{\hat{I}}{\pi} \left[1 + \frac{\pi}{2} \cos \omega t + \frac{2}{3} \cos 2\omega t - \frac{2}{15} \cos 4\omega t + \ldots \right]$$

Its load is a tuned circuit comprising an inductor $L = 50\ \mu\text{H}$ having an intrinsic Q-factor $Q = 160$, in parallel with a capacitor $C = 50$ pF. When a load is coupled to this circuit in such a way that the tuning is unaltered, the Q-factor of L is reduced to an effective value $Q_e = 40$, and the measured dc component of the current traversing the circuit is 100 mA.

Find the rms voltage across the tuned-circuit at the fundamental frequency ω to which it is tuned, and the powers absorbed by the load and dissipated in the tuned-circuit, respectively, at ω. Estimate the voltage across the tuned circuit at 2ω.

13*

Interpretation

The calculations are done easily in steps, in terms of very close approximations for the admittance of the tuned circuit. The dc component of current yields the peak values of the ac components, for $0.1 = \hat{I}/\pi$, or $\hat{I} = 0.1\pi$.

(1) Load absent: $G = \dfrac{1}{Q\omega L} = \dfrac{1}{Q}\sqrt{\dfrac{C}{L}} = \dfrac{10^{-3}}{160}\,\text{S}$

(2) Load present: $G_e = \dfrac{1}{Q_e\omega L} = \dfrac{1}{Q_e}\sqrt{\dfrac{C}{L}} = \dfrac{10^{-3}}{40}\,\text{S}$

(3) Contribution from load $= G_L = G_e - G = 3\times10^{-3}/160\,\text{S}$

(4) The fundamental-frequency voltage is

$$\hat{V}_\omega = \hat{I}_\omega/2G_e = 2\pi\times10^3, \quad \text{or} \quad V_\omega = 2\pi\times10^3/\sqrt{2}\,\text{V, rms}$$

(5) Power absorbed by load $= G_L V_\omega^2 = 3\pi^2\times10^3/80\,\text{W}$

Power dissipated in circuit $= G V_\omega^2 = \pi^2\times10^3/80\,\text{W}$.

(6) At the second harmonic frequency 2ω the susceptance of the tuned circuit may be assumed to overwhelm the conductance, which can therefore be ignored. Thus,

$$|Y_{2\omega}| = 2\omega C - 1/2\omega L = 2\sqrt{\dfrac{C}{L}} - \dfrac{1}{2}\sqrt{\dfrac{C}{L}} = \dfrac{3}{2}\times10^{-3}\,\text{S}$$

and therefore

$$V_{2\omega} = I_{2\omega}/|Y_{2\omega}| = \dfrac{0.1\pi\times2}{3\pi\sqrt{2}}\cdot\dfrac{2\times10^3}{3} = \dfrac{400}{9\sqrt{2}}\,\text{V, rms}$$

Comment

The operating condition is Class B, in which the output current for a sinusoidal input is in the form of half-sine pulses. Nevertheless, the output voltage has a substantially pure sine waveform through the filtering action of the tuned circuit: $V_{2\omega}/V_\omega = 1/45\pi$, approximately, and the second harmonic is therefore less than 1% of the fundamental.

Note the neatness of the form $G = \sqrt{(C/L)}/Q$, which obviates evaluation of ω.

ILLUSTRATION 5.9

A full-wave rectifier of negligible internal resistance delivers a voltage

$$e(t) = 100 + 66 \cos 628t - 13 \cos 1256t$$

The rectifier is connected, in series with a dc moving-coil ammeter and a thermocouple ammeter, to a load comprising a resistor of $2000\,\Omega$ in parallel with a capacitor of $2\,\mu F$. Calculate,

(a) the readings of the two ammeters,
(b) the rms voltage across the load,
(c) the power absorbed by the load,
(d) the power-factor of the load.

Interpretation

Since admittance calculations at two frequencies are required, there is some advantage in scaling the data.

Scaled to 1 V and $\omega = 1$ rad/s,

$$e(t) = 1 + 0.66 \cos t - 0.13 \cos 2t$$

where the voltage scaling factor is $K_v = 100$ and the frequency scaling factor is $\beta = 628$.

For resistance scaled to $R = 1\,\Omega$ the magnitude scaling factor is $\alpha = 2000/R = 2000$, and the scaled capacitance is $C = 2 \times 10^{-6} \alpha\beta = 2.512\,F$. The scaled conductance is $G = 1/R = 1$ S.

(a) The moving-coil meter reads the mean current. This is governed by the unidirectional component of voltage only, and

$$I_{\text{mean}} = \tfrac{1}{1} = 1\ A, \quad \text{scaled}$$

The thermocouple meter reads the rms current, which is found from the instantaneous current. This has the form

$$i(t) = E_0 G + \hat{E}_1 |Y_1| \cos(\omega t + \phi_1) - \hat{E}_2 |Y_2| \cos(2\omega t + \phi_2)$$
$$= I_0 + \hat{I}_1 \cos(\omega t + \phi_1) - \hat{I}_2 \cos(2\omega t + \phi_2)$$

where in terms of the scaled values,

$$I_0 = 1\ A$$
$$I_1 = 0.66\sqrt{\{G^2 + \omega^2 C^2\}} = 0.66\sqrt{\{1 + (2.512)^2\}} = 1.78\ A$$
$$I_2 = 0.13\sqrt{\{G^2 + (2\omega C)^2\}} = 0.13\sqrt{\{1 + (5.024)^2\}} = 0.665\ A$$

The scaled rms current is then

$$I = \sqrt{\left\{ I_0^2 + \frac{\hat{I}_1^2 + \hat{I}_2^2}{2} \right\}} = 1.675\ A, \quad \text{scaled}$$

(b) The rms voltage across the load is the rms value of the rectifier output voltage, which is

$$V = \sqrt{\left\{ E_0^2 + \frac{\hat{E}_1^2 + \hat{E}_2^2}{2} \right\}} = 1\cdot104 \text{ V}, \quad \text{scaled}$$

(c) The power absorbed by the load is simply

$$P = V^2/R = 1\cdot22 \text{ W}, \quad \text{scaled}$$

(d) Power-fractor $= \dfrac{\text{True power}}{\text{Apparent power}} = \dfrac{P}{V_{\text{rms}}I_{\text{rms}}} = 0\cdot66.$

It is now simple to rescale these results for the actual circuit values. Multiplying results (a) by $K_v/\alpha = 0\cdot05$, (b) by $K_v = 100$, (c) by $K_v^2/\alpha = 5\cdot0$ and leaving (d) unchanged,

(a) $I_{\text{mean}} = 1\times0\cdot05 = 0\cdot05$ A, actual
    $I_{\text{rms}} = 1\cdot675\times0\cdot05 = 0\cdot0838$ A, actual
(b) $V_{\text{rms}} = 1\cdot104\times100 = 110\cdot4$ V, actual
(c) $\quad P = 1\cdot22\times5\cdot0 = 6\cdot1$ W, actual
(d) Power-factor $= 0\cdot65$, actual or scaled.

Comment

(1) This illustration demonstrates the applicability of scaling when more than one frequency component is involved. Note the simplicity of the calculations, and the ease of scaling in both directions.

(2) Note that in this case of a resistor in parallel with a capacitor, the power is given directly by V_{rms}^2/R: the capacitor absorbs no mean power. The result, $P = V_{\text{rms}}^2/R$, is equivalent to the more tedious calculation $P = V_0 I_0 + V_1 I_1 \cos\phi_1 + V_2 I_2 \cos\phi_2$ when ϕ_1 and ϕ_2 are the arguments of the load admittance at ω and 2ω.

ILLUSTRATION 5.10

An emf $e(t) = 60 + 100 \sin \omega t + 30 \sin (3\omega t + \pi/3)$, where $\omega = 314$, is applied to a series circuit comprising a resistor of $120\,\Omega$, a capacitor of $17\cdot7\ \mu\text{F}$ and an inductor of $63\cdot6$ mH.

Obtain an expression for the steady-state current, and calculate:
(a) the rms current,
(b) the power dissipated in the circuit, and
(c) the rms voltage across the capacitor.

(L.U. Part 2, Electrical Theory and Measurements).

Interpretation

The steady-state impedance of the circuit at any angular frequency $n\omega$ is

$$Z_{n\omega} = \sqrt{\{R^2 + (n\omega L - 1/n\omega C)^2\}}\underline{/\theta_{n\omega}}$$

For $n = 1$ (fundamental),

$$Z_{\omega} = 200\underline{/-53°}$$

For $n = 3$,

$$Z_{3\omega} = 120\underline{/0}$$

There is no dc component in the steady-state current owing to the capacitor in series, and thus

$$i(t) = \tfrac{100}{200} \sin (\omega t + 53\cdot1°) + \tfrac{30}{120} \sin (3\omega t + \pi/3)$$

and

$$I_{rms} = \sqrt{\left\{\frac{(0\cdot5)^2 + (0\cdot25)^2}{2}\right\}} = 0\cdot395 \text{ A}$$

The power dissipated in the circuit resides wholly in the resistor, and is thus simply

$$P = P_R = I_{rms}^2 R = 18\cdot7 \text{ W}$$

The voltage across the capacitor is the difference between $e(t)$ and the complex volt-drop across R and L. This, however, is difficult to evaluate, for it involves combinations of voltages in different phases. The alternative, which is much simpler, is to use the principle of superposition. If the ac components did not exist in the supply, the capacitor would obviously become charged to the steady value of 60 V, which is the dc component in the supply. When the ac components are included, their contributions of voltage-drop are merely superimposed on this dc voltage. Thus, symbolically,

$$v_c(t) = V_0 + \frac{\hat{I}_\omega \sin (\omega t + \emptyset_\omega - \pi/2)}{\omega C} + \frac{\hat{I}_{3\omega} \sin (3\omega t + \pi/3 - \pi/2)}{3\omega C}$$

and substituting values,

$$v_c(t) = 60 + 90 \sin (\omega t - 37°) + 15 \sin (3\omega t - 30°)$$

The rms voltage, which is unaffected by the phases, is then

$$V_{rms} = \sqrt{\left\{(60)^2 + \frac{(90)^2 + (15)^2}{2}\right\}} = 88 \text{ V}$$

Comment

(1) A unidirectional component of voltage exists across the capacitor, although there is no steady-state dc component of current.

(2) The superposition principle affords an easy approach to the voltage calculation, although the unidirectional component is predicted intuitively.

(3) Note the position of the dc term in the rms voltage formula: this is a case in which it is not valid to regard dc as zero-frequency ac.

CHAPTER 6

Electronic amplifiers with feedback circuits

INTRODUCTION

Circuits or systems incorporating reverse transmission or feedback paths are generally regarded as distinctive and endowed with special properties. While feedback is fundamental to some systems, such as error-actuated servo-systems and common types of oscillator, to others, such as amplifiers, it is not fundamental to their action but is an addition for the modification of intrinsic behaviour. Even oscillators do not all depend on feedback to provide a self-sustaining signal, an exception being the negative-conductance oscillator, realisable with a tunnel diode. It is noteworthy that feedback oscillators can be viewed equivalently in terms of negative admittance (see Illustrations 6.7 and 6.8).

In respect of network analysis, there is nothing exclusive about an amplifier with feedback circuitry: each of the many possible arrangements is subject to the normal procedures of nodal-voltage or loop-current analysis, and to matrix manipulation. The specification of a two-port network or amplifier by a matrix in fact automatically embraces feedback in its reverse parameters, such as Y_{12}, h_{12}. The intrinsic feedback in a transistor, an imperfection often requiring neutralisation in high-frequency amplifiers, is reflected in h_{re} ($\equiv h_{12}$) or by $C_{b'c}$ and $g_{b'c}$ in the corresponding hybrid-π circuit model.

Many active circuits may be viewed either as networks incorporating amplifiers or as amplifiers with feedback circuitry. But when the circuitry external to the amplifier is at all complicated, no simple feedback formulae can be invoked readily, and direct analysis is appropriate. The cases of composite feedback in Illustrations 6.12 and 6.13 and the active filter in Illustration 6.14 are examples. Nevertheless, there are traditional yet common feedback arrangements, notably cases (a) and (b) in Illustration 6.1, for which simple general formulae are applicable, such as $A_f = A/(1+A\beta)$, provided the assumptions on which the formulae are based are satisfied adequately (care is necessary with junction transistors). Such formulae are of great value, since they display the characteristics of the amplifier explicitly in A and those of the feedback network in β. The limitations are considered in Illustration 6.4, and comparison is made with a two-port matrix approach in Illustration 6.5.

ILLUSTRATION 6.1

What are the basic arrangements for electronic feedback systems and what are the equations and matrices governing them?

Interpretation

The arrangements of amplifier A and feedback network β are shown in Figure 6.1.

(a) Voltage-dependent series-applied voltage feedback

This is the common conception of a feedback system, attributed to H. S. Black ('Stabilised Feedback Amplifiers,' Bell System Technical Journal, January 1934). Note that the polarity convention in this paper

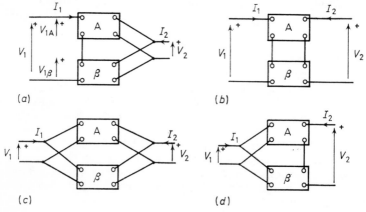

(a) (b) (c) (d)

Figure 6.1

is opposite to that which is now preferred, having an electro-mechanical analogue in an error-acuated servo-system. Let A be the intrinsic amplifier voltage gain, and let the input impedance be great and the output impedance be low so that I_1 and the output loading by the voltage-dividing β-network are negligible. Then,

$$V_{1A} = V_1 - \beta V_2, \quad V_2 = AV_{1A} = AV_1 - A\beta V_2$$

whence

$$V_2/V_1 = A/(1+A\beta) \tag{1}$$

If the stated impedance conditions are not applicable, which may be so with junction transistors, equation (1) is an approximation. But as the configuration conforms to h-matrix addition, all loading effects may be accounted for in the alternative equation

$$[h] = [h]_A + [h]_\beta \tag{2}$$

See Illustrations 1.15 and 6.5.

(b) Current-dependent series-applied voltage feedback

In this arrangement (formerly misnamed current feedback) the voltage fed back is proportional to the current output into the load. Adaption to equation (1) is possible ($\beta \simeq R_f/(R_f + R_2)$ for a feedback resistor R_f in series with a load R_2), but the individual circuit analysis is preferred. In general the configuration conforms to Z-matrix addition, and exact solutions may be obtained from

$$[Z] = [Z]_A + [Z]_\beta \tag{3}$$

See Illustrations 1.18 and 6.12.

(c) Voltage-dependent parallel-applied current feedback

This is the dual of (b) and in general

$$[Y] = [Y]_A + [Y]_\beta \tag{4}$$

The common example is a two-terminal admittance bridged across the high-potential input and output terminals of the amplifier. See Illustrations 1.9, 1.11, 6.6 and 6.8. The neutralisation of a transistor, with the object of cancelling Y_{12}, is approached in this way.

(d) Current-dependent parallel-applied current feedback

This is the inverse of (a) and is solvable exactly by combining inverted h-matrices. Where the h-matrix equation has the form

$$\begin{bmatrix} h_{11} & h_{21} \\ h_{12} & h_{22} \end{bmatrix} \cdot \begin{bmatrix} I_1 \\ V_2 \end{bmatrix} = \begin{bmatrix} V_1 \\ I_2 \end{bmatrix}$$

its inverse has the form

$$\begin{bmatrix} g_{11} & g_{21} \\ g_{12} & g_{22} \end{bmatrix} \cdot \begin{bmatrix} V_1 \\ I_2 \end{bmatrix} = \begin{bmatrix} I_1 \\ V_2 \end{bmatrix}$$

and (subject to validity tests) the g-parameters for arrangement (d) add as the h-parameters do for arrangement (a), or

$$[g] = [g]_A + [g]_\beta \tag{4}$$

Given $[h]$,

$$[g] = [h]^{-1} = \frac{1}{\Delta_h} \begin{bmatrix} h_{22} & -h_{12} \\ -h_{21} & h_{11} \end{bmatrix} \tag{5}$$

Comment

Arrangement (a) is very common, both as the basis of negative feedback in amplifiers and for positive feedback in oscillators. In many negative feedback amplifier applications the β network is simply a resistance potential divider, and the isolating transformer necessary in Figure 6.6, Illustration 6.5, can often be excluded by appropriate circuitry (as, for example, in second collector to first emitter feedback in a two-stage CE transistor amplifier).

The usual case of arrangement (b), realising negative feedback, is no more than an unbypassed resistor in series with the emitter in a CE transistor circuit (or source in an FET, or cathode in a valve circuit). See Illustration 1.18.

Arrangement (c), applied in the form mentioned to a polarity-reversing amplifer, realises negative feedback. It exhibits low input impedance (by contrast with its dual in Illustration 1.18), and with an additional series admittance is a potential operational amplifier. This class of amplifier is now very important, not only for its analogue applications, but also as the basis of a wide range of active RC filters.

Feedback affects not only gain but also impedances. Arrangements (a) and (b) produce similar effects on gain but opposite effects on output impedance. A hybrid arrangement (called *bridge feedback* in the valve era) may be used to control output impedance independently of gain. See Illustration 6.12.

ILLUSTRATION 6.2

In a feedback system the parameters of both amplifier and feedback network are in general complex functions of frequency.

(a) What criterion distinguishes negative feedback from positive feedback when the system parameters are complex quantities?

(b) What is the *Nyquist criterion* as applied to arrangement (a) in Illustration 6.1?

(c) Define a stable network or system, and compare equivalent ways of viewing stability and the threshold of oscillation.

Interpretation

(a) The criterion is the effect on gain-magnitude. Feedback is negative or degenerative when $|A_f| < |A|$ and positive or regenerative when $|A_f| > |A|$. For voltage-dependent series applied voltage feedback (arrangement (a) in Illustration 6.1), $A_f = A/(1 + A\beta)$. The feedback is negative when $|1 + A\beta| > 1$, and when $A\beta = x + jy$, $|1 + A\beta| > 1$ when $(1 + x)^2 + y^2 > 1$.

(b) The boundary between negative and positive feedback is set by $|1 + A\beta| = 1$. The non-trivial condition for this when

$$(1 + x)^2 + y^2 = 1$$

or

$$x^2 + 2x + y^2 = 0$$

This is the equation for a circle of unit radius centered at the point $-1, 0$ in the complex plane, and is the locus of $A\beta$ for which the gain magnitude is unaffected by the feedback.

$A\beta$ is represented by phasors in Fig. 6.2(a) for $A\beta = 1 \cdot 5\underline{/150°}$ at ω_1 and $2 \cdot 0\underline{/240°}$ at ω_2. $|1 + A\beta|$ is 0·807 at ω_1 and 1·73 at ω_2, and the feedback is therefore regenerative for $A\beta$ within the circle but degenerative for it outside. The condition becomes critical as $A\beta \rightarrow -1\underline{/0}$; for then $A_f \rightarrow \infty$ and the system becomes unstable, in the sense that

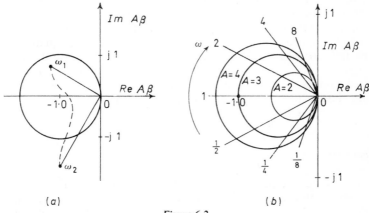

Figure 6.2

its output may not vanish as its input becomes vanishingly small. The phasors for ω_1 and ω_2 alone do not, therefore, confirm the stability of the system; for the point -1, 0 is likely to be enclosed by the locus of $A\beta$ for frequencies between ω_1 and ω_2, unless it makes a detour as indicated by the dotted path: such a locus defines a state that is *conditionally stable*, for circuit changes may dause the point -1, 0 to be enclosed.

The criterion of H. Nyquist ('Regeneration Theory', Bell System Technical Journal, January 1932) for a feedback amplifier or system of the kind under consideration to be stable, or at least conditionally so, is that the point -1, 0 in the complex-plane representation of $A\beta$ shall not be encircled by the locus of $A\beta$ as a function of frequency from $\omega = 0$ to $\omega = \infty$. If the point is encircled, the system is unstable. The criterion is not simple to prove formally as a generality; but it can be reasoned intuitively from the critical condition $A\beta = -1$, for which a supposed input signal is just returned through the β-path, so that if the loop A, β is closed the signal is just self-sustaining and the system is at the threshold of action as a self-ocillator or generator. If $A\beta$ is given a real increment δ so that $A\beta = -(1+\delta)$, that point on its new locus which is the intersection of that locus with the negative real axis must be δ to the left of the point -1, 0, and the new locus may therefore be said to encircle this point. But when $A\beta = -(1+\delta)$ the returned signal exceeds the initiating one, which is therefore reinforced rather than merely sustained. Thus, if $A\beta = -1$ is the threshold of oscillation or instability, $A\beta = -(1+\delta)$ is beyond it and oscillation with growing amplitude, constrained in practice by non-linearity, must exist.

(c) A *stable network* is one whose transient response decays with time. Equivalently, it is one whose response to a vanishingly small excitation is vanishingly small. Figure 6.2(b) shows the loci of $A\beta$ for Figure 6.3, Illustration 6.3. For $A < 3$ the positive feedback merely increases the gain (Table 2, Illustration 6.3); but when $A = 3$ the locus passes through -1, 0, and this value of A is shown in Illustration 6.3 to be that for which the transient response (given by equation (3)) just fails to decay and the natural frequencies just move on to the imaginary axis in the complex-frequency plane. The locus shown for $A = 4$, implies positive real parts for the natural frequencies and growth with time in the transient response. It could never exist in the form shown, in practice.

Comment

The Nyquist criterion applies directly to feedback amplifiers or systems conforming to Figure 6.1(a), and is not readily adapted to other arrangements. By contrast, the requirement for stability that transient response must decay, or, equivalently, that the natural frequencies must have negative real parts, is independent of circuit or system arrangement.

ILLUSTRATION 6.3

Figure 6.3 is an amplifier employing the same circuit elements as Figure 2.8, Illustration 2.10, but excited instead from a voltage-source $v_1(t)$ and with the circuit C_1, R_1, C_2, R_2 re-arranged to provide voltage-dependent series-applied voltage feedback. $C_1 = C_2 = 1$ F and $R_1 = R_2 = 1\ \Omega$ as before.

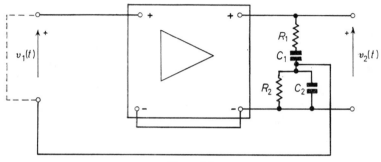

Figure 6.3

(1) Determine the natural frequencies or poles of $v_2(t)$ for the closed-loop condition when $A = 1, 2, 3$ and 4. Locate them on a complex-frequency plane diagram, and consider its implications in respect of the stability of the system.

(2) Express the steady-state closed-loop gain for sinusoidal excitation as a function of A and ω for the given circuit values. What are the gains at $\omega = 1$ when $A = 1, 2, 2·5$ and 3?

Interpretation

(1) When $v_1(t)$ is reduced to zero but with its path closed as indicated by the dotted line, the circuit is exactly the same as Figure 2.8 when $i_1(t)$ is reduced to zero but with its path open. Hence, for $C_1 = C_2 = C$ and $1/R = G$, where $R = R_1 = R_2$, the homogeneous or excitation-free equation for $v_2(t)$, which is $Av_1(t)$ in Figure 2.9, has exactly the same form as equation (3) in Illustration 2.10; or,

$$\{D^2 + [(3 - A)G/C]D + G^2/C^2\}v_2(t) = 0 \tag{1}$$

The corresponding characteristic equation, equation (4), Illustration 2.10,

$$s^2 + [(3-A)G/C]s + G^2/C^2 = 0 \tag{2}$$

gives the natural frequencies in Table 6.1. They are shown as poles on the complex-frequency plane diagram of Figure 6.4.

Table 6.1

A	s_1	s_2
1	$-1 \cdot 0$	$-1 \cdot 0$
2	$-\frac{1}{2} + j\sqrt{3}/2$	$-\frac{1}{2} - j\sqrt{3}/2$
3	$j1 \cdot 0$	$-j1 \cdot 0$
4	$\frac{1}{2} + j\sqrt{3}/2$	$\frac{1}{2} - j\sqrt{3}/2$

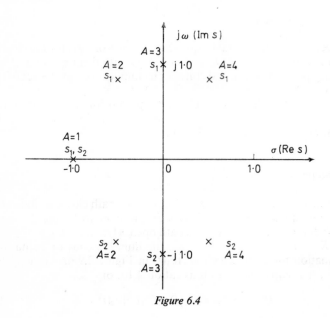

Figure 6.4

The response of the amplifier to an input $U(t)v_1(t)$ would have transient and steady-state components in the form

$$v_2(t) = v_{2tr} + v_{2ss} = Ae^{s_1 t} + Be^{s_2 t} + v_{2ss}$$

$$= e^{\sigma t}(Ae^{j\omega t} + Be^{-j\omega t}) + v_{2ss} \tag{3}$$

$$= Ke^{\sigma t} \sin(\omega t + \phi) + v_{2ss}$$

The stability requirement for the system is that the transient component in the response must die out. Equation (3) shows that this can be so only when the real part σ of the complex natural frequencies is negative and the poles in Figure 6.4 are constrained to the left-half of the complex-frequency plane. The threshold of instability is reached when $\sigma = 0$ and the poles are just on the imaginary axis, as when $A = 3$; for then the transient component is just self-sustaining. When $A > 3$, σ becomes positive, as exemplified by the case $A = 4$, and indefinite growth in amplitude is implied; but in practice this growth is retarded and σ is diminished as the non-linear region of amplifier operation is entered, so that the system stabilises itself, at the expense of waveform distortion, at some finite amplitude of self-oscillation.

(2) For Figure 6.3 under steady sinusoidal, the closed-loop gain has the general form

$$A_f = V_2/V_1 = A/(1 + A\beta) \tag{4}$$

The complex feedback factor β is derived for an equivalent arrangement in Illustration 6.7, and when $C_1 = C_2 = C$ and $R_1 = R_2 = R$, it may be put in the form

$$\beta = -1 \left/ \left[3 + j \left(\frac{\omega}{\omega_0} - \frac{\omega_0}{\omega} \right) \right] \right. \tag{5}$$

where $\omega_0 = 1/CR$. Putting $CR = 1$ for this case and substituting in equation (4) then gives

$$A_f = \frac{A}{1 - A/[3 + j(\omega - 1/\omega)]} \tag{6}$$

whence the squared magnitude is

$$|A_f|^2 = \frac{A^2(\omega^4 + 7\omega^2 + 1)}{\omega^4 + (A^2 - 6A + 7)\omega^2 + 1} \tag{7}$$

The gains are shown in Table 6.2 for the three values A and ω.

14

Table 6.2

ω	$\|A_f\|$		
	$A = 1$	$A = 2$	$A = 2 \cdot 9$
$\frac{1}{2}$	1·34	3·72	6·47
1	1·5	6	87
2	1·34	3·72	6·47

Comment

The imaginary term in equation (5) bears comparison with the normalised response for a tuned circuit (see Illustration 4.8), and Table 6.2 indicates that the loop-gain is frequency selective like a tuned circuit. As $A \to 3$, $A_f \to \infty$ and the threshold of oscillation is approached. This is consistent with σ vanishing from equation (3) and the poles falling onto the imaginary axis in Figure 6.4.

ILLUSTRATION 6.4

The voltage gain of an amplifier with voltage-dependent series-applied voltage feedback is commonly accepted as $A/(1+A\beta)$, where A is the intrinsic gain of the amplifier and β is the voltage-feedback factor. What are the requirements for this expression to be accurate, in what respects may they be departed from in practice, and what are the attendant modifications to the closed and open-loop gains?

Interpretation

The given formula, derived in Illustration 6.1, depends on the assumptions (1), that the β-network does not alter the intrinsic gain A of the amplifier by loading its output terminals appreciably; and (2), that the amplifier input impedance is infinite and $I_1 = 0$ so that the β-network is operating under open-circuit conditions and the amplifier input is truly $V_1 - \beta V_2$. When assumptions (1) and (2) are valid, it is also true for the arrangement of Figure 6.5(a) that

$$A_f = \frac{V_2}{V_1} = \frac{V_2}{E_1} = \frac{A}{1+A\beta}, \quad \text{exactly} \tag{1}$$

since under the idealised conditions there is no volt-drop in Z_1.

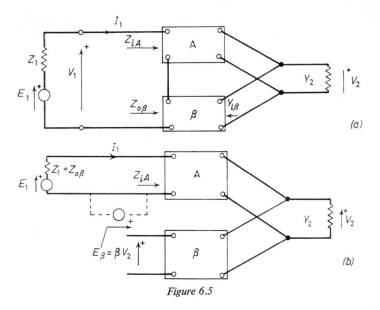

Figure 6.5

In Figure 6.5(a), assumption (1) is not valid when $Y_{i\beta}$ is not negligible compared with the load Y_2. However, it is simple to merge $Y_{i\beta}$ with Y_2 when calculating A for a given load and set of amplifier parameters.

Assumption (2) is not valid when Z_{iA} is not high, compared with $Z_{0\beta}$ alone in respect of V_2/V_1, and compared with Z_1 additionally, in respect of the true closed and open-loop gains. Let the β-network be represented, in respect of its action in the amplifier input circuit, by an equivalent generator of impedance $Z_{0\beta}$ and emf $E_\beta = \beta V_2$, where βV_2 is the open-circuit voltage of the β-network. Let this be considered to act first in series with the input generator terminal voltage V_1. Then,

$$V_2 = AZ_{iA}I_1 = \frac{AZ_{iA}(V_1 - E_\beta)}{Z_{iA} + Z_{0\beta}} = \frac{AZ_{iA}(V_1 - \beta V_2)}{Z_{iA} + Z_{0\beta}}$$

whence the gain referred to V_1 is

$$A_f = \frac{V_2}{V_1} = \frac{A}{1 + \left[\dfrac{Z_{0\beta}}{Z_{iA}} + A\beta\right]} \qquad (2)$$

Equation (2) expresses the modified gain in practical form, from input to output terminals. But the true closed loop is through E_1 and Z_1; and as Z_1 forms with Z_{iA} a potential divider augmenting the β-network, it cannot be excluded in the context of stability criteria: both closed and open-loop gains must therefore be related to E_1 rather than

V_1, except under the ideal conditions ($Z_{iA} = \infty$, $I_1 = 0$) implied in equation (1). Let $Z_{0\beta} + Z_1 + Z_{iA} = Z_T$. Then, representing the β-network by an equivalent generator as before,

$$V_2 = AZ_{iA}I_1 = AZ_{iA}(E_1 - \beta V_2)/Z_T$$

whence the closed-loop gain relative to E_1 is

$$A_{CL} = \frac{V_2}{E_1} = \frac{AZ_{iA}/Z_T}{1 + (AZ_{iA}/Z_T)\beta} \tag{3}$$

This can be set in the pattern of equation (1) as

$$A_{CL} = A_e/(1 + A_e\beta) \tag{4}$$

where $A_e = AZ_{iA}/Z_T$ is the effective amplifier gain relative to E_1 when the emf of the feedback source is suppressed but its impedance $Z_{0\beta}$ is included. The open-loop gain, corresponding to but not equal to $A\beta$ in equation (1), is thus

$$A_{CL} = A_e\beta = A\beta Z_{iA}/Z_T \tag{5}$$

The open-loop gain can be calculated or measured with the system arranged as in Figure 6.5(b). $Z_{0\beta}$ is included in the amplifier input loop, for E_1, and E_β when it acts in the loop, both drive their contributions of current through it. E_β is the calculable or measurable open-circuit voltage βV_2 of the disconnected β-network as indicated. From Figure 6.5(b), $\beta V_2 = AE_1Z_{iA}\beta/Z_T$; and when $\beta V_2 = -E_1$, $AZ_{iA}\beta/Z_T = -1$, which is consistent with the threshold of oscillation given by equation (4) in the form $1 + A_e\beta = 0$, or $A_e\beta = -1$.

Comment

This illustration has shown the limitations of equation (1), which applies to the idealisation of a common feedback system, and has distinguished terminal-voltage gain from true closed-loop gain as it applies when the idealisation is not valid. The corrections indicated are often important with junction-transistor amplifiers, for the input impedance of a junction transistor can be quite low.

The gains of other feedback arrangements, such as Figure 6.1(b) and (c) can also be expressed in terms of a feedback factor and adapted to the style of equation (1). In the author's opinion, however, such adaptations are clumsy and have little point as the arrangements are solvable elegantly and without approximations by combining matrices. See Illustrations 6.5, 6.6, 6.10, 6.12, and 1.18.

ILLUSTRATION 6.5

In Figure 6.6, the amplifier bounded by terminals a, b, c, d is of the junction transistor type and is specified by the matrix

$$[h]_{CE} = \begin{bmatrix} h_{ie} & h_{re} \\ h_{fe} & h_{oe} \end{bmatrix} = \begin{bmatrix} 1 \cdot 1 \times 10^3 & 2 \cdot 5 \times 10^{-4} \\ 50 & 2 \cdot 5 \times 10^{-5} \end{bmatrix}$$

Terminals b and d are internally connected, and the isolating transformer T steps down the feedback voltage from P to S in the ratio $2 : 1$,

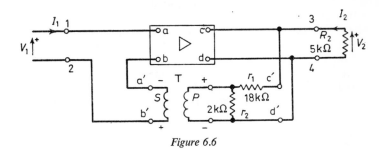

Figure 6.6

reversing the polarity as indicated. Calculate accurately the voltage gain and input impedance (a), for the amplifier and load R_2 alone; and (b), for the complete amplifier-feedback system. The transformer may be assumed ideal.

Interpretation

(a) The voltage gain of the amplifier with V_1 applied to a, b, and R_2 connected to c, d but with r_1, r_2 and T disconnected is given by manipulating the hybrid equations

$$h_{ie}I_1 + h_{re}V_2 = V_1$$

$$h_{fe}I_1 + h_{oe}V_2 = I_2 = -V_2/R_2$$

from which

$$A = \frac{V_2}{V_1} = h_{fe}/[h_{re}h_{fe} - h_{ie}(h_{oe} + 1/R_2)] \tag{1}$$

$$= -213$$

and

$$Z_{in} = \frac{V_1}{I_1} = h_{ie} - [h_{re}h_{fe}/(h_{oe}+1/R_2)] \tag{2}$$

$$= 1 \cdot 04 \text{ k}\Omega$$

(b) The transistor specification with h-parameters is ideal for solving the complete feedback system, for it is a unique property of h-matrices that for two networks connected in series on the input side and in parallel on the output, the system matrix is simply the sum of the h-matrices for its constituents. This is subject to a validity criterion which in this case is satisfied by inclusion of the isolating transformer T (see Illustration 1.15).

The h-matrix for the feedback network alone, bounded by a', b', c', d', is almost evident by inspection. Let the impedance ratio of T in the direction P → S be λ^2, where λ, the voltage ratio, is $-1/2$. The constraints $V_2 = 0$, $I_1 = 0$ imposed on the equations $h_{11}I_1 + h_{12}V_2 = V_1$, $h_{21}I_1 + h_{22}V_2 = I_2$ are equivalent to imposing short and open-circuit conditions on either the output or the input side, so that

$$h_{11} = \frac{V_1}{I_1}\bigg|_{V_2 = 0} = Z^{sc}_{a'-b'} = \frac{r_1 r_2 \lambda^2}{r_1 + r_2} = 450 \ \Omega$$

$$h_{22} = \frac{I_2}{V_2}\bigg|_{I_1 = 0} = Y^{oc}_{c'-d'} = \frac{1}{r_1 + r_2} = 5 \times 10^{-5} \text{ S}$$

$$h_{12} = \frac{V_1}{V_2}\bigg|_{I_1 = 0} = \frac{V_2 r_2 \lambda}{(r_1 + r_2)V_2} = -\frac{1}{20}$$

and, using the property of a reciprocal network,

$$h_{21} = -h_{12} = \tfrac{1}{20}$$

Thus,

$$[h]_f = \begin{bmatrix} 450 & -0 \cdot 05 \\ 0 \cdot 05 & 5 \times 10^{-5} \end{bmatrix}$$

and the total matrix is therefore

$$[h] = [h]_{CE} + [h]_f = \begin{bmatrix} 1 \cdot 55 \times 10^3 & -4 \cdot 98 \times 10^{-2} \\ 50 \cdot 1 & 7 \cdot 5 \times 10^{-5} \end{bmatrix}$$

Adapting the matrix elements to equations (1) and (2), the voltage gain A_f and the input impedance $Z_{in(f)}$, referred to terminals 1, 2, are

$$A_f = -17 \cdot 2 \text{ and } Z_{in(f)} = 10 \cdot 6 \text{ k}\Omega$$

Comment

(1) It is instructive to compare the above results with those that may be calculated from $A_f = A/(1+A\beta)$ and $Z_{in(f)} = Z_{in}(1+A\beta)$. In this case, $\beta = h_{12} = -1/20$, $A = -213$, and $Z_{in} = 1.04$ kΩ. Then,

$$A_f = A/(1+A\beta) = -213/11.7 = -18.2$$

and

$$Z_{in(f)} = Z_{in}(1+A\beta) = 1.04 \times 11.7 = 12.2 \text{ k}\Omega$$

Compared with the correct values calculated by the matrix method, the gain is 6% high and the impedance 15% high. These discrepancies are accounted for by the failure of the simple feedback formula to allow for full loading effects (see Illustration 6.4).

(2) The matrix approach poignantly demonstrates the linearising and stabilising effects of negative feedback. In the summation of matrices for an amplifier and feedback system, non-linearities and fluctuations in the amplifier matrix (due, for example, to unstable operating conditions), may be swamped by the linear, stable elements in the matrix for a linear, passive feedback network, to yield a composite matrix of improved linearity and stability. This is equivalent to the statement $A_f \rightarrow -1/\beta$ for $A\beta$ great, but is comprehensive since the influence of each of the four matrix elements may be separately discerned.

(3) Note that the feedback formula $A_f = A/(1+A\beta)$ is consistent with the polarity conventions for two-port matrices.

(3) A network is said to be stable if its output in response to an external input vanishes as that input becomes vanishingly small. The threshold of stability is approached as the gain tends to infinity, as when $1+A\beta \rightarrow 0$ in the expression $A_f = A/(1+A\beta)$. This is also the threshold of oscillation when the feedback loop is closed. An equivalent criterion is readily stated in terms of the net h-matrix. In general symbols,

$$A_f = h_{21}/[h_{12}h_{21}-h_{11}(h_{22}+1/R_2)]$$

and $A_f \rightarrow \infty$ as

$$h_{12}h_{21}-h_{11}(h_{22}+1/R_2) \rightarrow 0$$

which is therefore a criterion for oscillation.

ILLUSTRATION 6.6

A common-emitter junction-transistor amplifier having input terminals 1, 2 and output terminals 3, 4 is specified by the matrix

$$\begin{bmatrix} h_{ie} & h_{re} \\ h_{fe} & h_{oe} \end{bmatrix} = \begin{bmatrix} 1.1 \times 10^3 & 2.5 \times 10^{-4} \\ 50 & 2.5 \times 10^{-5} \end{bmatrix}$$

Terminals 2, 4 are common, and parallel feedback is applied by means of a resistor $R_f = 20\ \text{k}\Omega$ bridged across terminals 1, 3. Calculate rigorously the voltage gain and input impedance when the load at terminals 3, 4 is a resistor $R_2 = 5\ \text{k}\Omega$.

Interpretation

The h-parameters are not easily adapted to this feedback configuration. But it is simple to transform them into Y-parameters, which are applicable directly as the arrangement is essentially one of paralleled two-port networks.

The h and Y-parameter equations have the general forms

$$h_{11}I_1 + h_{12}V_2 = V_1$$
$$h_{21}I_1 + h_{22}V_2 = I_2 \tag{1}$$

and

$$Y_{11}V_1 + Y_{12}V_2 = I_1$$
$$Y_{21}V_1 + Y_{22}V_2 = I_2 \tag{2}$$

Rearranging equations (1) to match the pattern of (2) gives

$$\frac{1}{h_{11}} \cdot V_1 - \frac{h_{12}}{h_{11}} \cdot V_2 = I_1$$

$$\frac{h_{21}}{h_{11}} \cdot V_1 + \left(h_{22} - \frac{h_{21}h_{12}}{h_{11}}\right) V_2 = I_2$$

whence by comparing coefficients,

$$Y_{11} = 1/h_{11}, \quad Y_{12} = -h_{12}/h_{11}, \quad Y_{21} = h_{21}/h_{11},$$
$$Y_{22} = h_{22} - h_{21}h_{12}/h_{11}$$

Identifying the subscripts with the transistor notation and substituting numerical values then gives

$$[Y]_{CE} = \begin{bmatrix} Y_{ie} & Y_{re} \\ Y_{fe} & Y_{oe} \end{bmatrix} = \begin{bmatrix} 9 \cdot 09 \times 10^{-4} & -2 \cdot 27 \times 10^{-7} \\ 4 \cdot 55 \times 10^{-2} & 1 \cdot 36 \times 10^{-5} \end{bmatrix}$$

The feedback resistor R_f may be treated as the single series element of an elemental two-port network (see Illustration 1.13), for which

$$|Y_f| = \begin{bmatrix} 1/R_f & -1/R_f \\ -1/R_f & 1/R_f \end{bmatrix} = \begin{bmatrix} 5 \times 10^{-5} & -5 \times 10^{-5} \\ -5 \times 10^{-5} & 5 \times 10^{-5} \end{bmatrix}$$

The total matrix for the parallel-feedback amplifier is thus

$$[Y] = [Y]_{CE} + [Y]_f = \begin{bmatrix} 9 \cdot 59 \times 10^{-4} & -5 \cdot 02 \times 10^{-5} \\ 4 \cdot 55 \times 10^{-2} & 6 \cdot 36 \times 10^{-5} \end{bmatrix}$$

Introducing the load admittance $G_2 = 1/R_2 = 2 \times 10^{-4}$ S, the final nodal-voltage equations are

$$(9 \cdot 59 V_1 - 0 \cdot 502 V_2) 10^{-4} = I_1$$
$$(455 V_1 + 0 \cdot 636 V_2) 10^{-4} = I_2 = -2 \times 10^{-4} V_2$$

whence the voltage amplification is

$$A = V_2/V_1 = -173$$

and the input impedance is

$$Z_{in} = V_1/I_1 = 104 \, \Omega$$

Comment

While the feedback admittance $1/R_f$ swamps the reverse transfer admittance Y_{re}, it is negligible against the much larger forward one Y_{fe}. These effects are reflected in the relatively small reduction in gain, from -213 without R_f to -173 with it, a reduction of only 19%; and the large reduction in input impedance, which falls from $1 \cdot 04$ kΩ to $104 \, \Omega$. This is consistent with the sensitivity of the expression $Y_{in} = Y + Y_f(1-A)$ (see Illustration 1.9) to A. The reader may check that $Y_{in} = 1/104$ S, on inserting $Y = 9 \cdot 61 \times 10^{-4}$ S, $Y_f = 5 \times 10^{-5}$ S and $A = -173$.

ILLUSTRATION 6.7

A phase-shift oscillator of the type shown in Figure 6.7(a) is required to oscillate at an angular frequency ω. At this frequency, however, the amplifier has an appreciable phase-shift θ. If the complex gain of the amplifier is $A/\underline{\theta}$, find the equations to be satisfied for oscillation to be sustained at ω. The amplifier input impedance may be assumed to be infinite, and loading effects at its output terminals may be ignored.

In a particular case, $R_1 = R_2 = R$, $C_1 = C_2 = C$, $CR = 1$ and $\theta = 45°$. Find the magnitude of the amplifier gain at the threshold of oscillation and the angular frequency of oscillation.

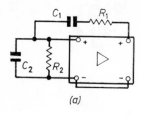

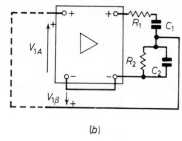

(a) *(b)*

Figure 6.7

Interpretation

Figure 6.7(a) is adapted to the form and polarity conventions of Figure 6.1(a), Illustration 6.1, by rearranging it as in Figure 6.7(b). In this $V_{1\beta}$ is negative relative to the polarities in Figure 6.1(a), so that β is negative and an open-loop gain $A\beta = -1$ can be realised to satisfy the condition $1 + A\beta = 0$ at the threshold of oscillation. The equation $A\beta = -1$ is to be satisfied at ω when A and β are complex. Treating the circuit C_1, R_1 and C_2, R_2 as a feedback potential divider,

$$\beta = -\frac{R_2/(1 + j\omega C_2 R_2)}{[R_2/(1 + j\omega C_2 R_2)] + R_1 + (1/j\omega C_1)}$$

or

$$\frac{1}{\beta} = -\left[1 + \frac{R_1}{R_2} + \frac{C_2}{C_1} + j\left(\omega C_2 R_1 - \frac{1}{\omega C_1 R_2}\right)\right] = -(x + jy)$$

At the threshold of oscillation, $A\beta = -1\underline{/0}$, or

$$\left|\frac{1}{\beta}\right| = |A| \quad \text{and} \quad \arg\frac{1}{\beta} = \arg A = \theta$$

Thus,

$$|A| = \left|\frac{1}{\beta}\right| = \sqrt{(x^2 + y^2)}$$

$$= \sqrt{\left\{\left[1 + \frac{R_1}{R_2} + \frac{C_2}{C_1}\right]^2 + \left[\omega C_2 R_1 - \frac{1}{\omega C_1 R_2}\right]^2\right\}}$$

and

$$\arg\frac{1}{\beta} = \theta = \tan^{-1}\left[\frac{\omega C_2 R_1 - (1/\omega C_1 R_2)}{1 + (R_1/R_2) + (C_2/C_1)}\right]$$

(b) When $R_1 = R_2 = R$ and $C_1 = C_2 = C$,

$$\tan \theta = \tan\left(\arg \frac{1}{\beta}\right) = \frac{(\omega CR)^2 - 1}{3\omega CR}$$

and

$$|A| = \sqrt{\left\{9 + \left[\frac{(\omega CR)^2 - 1}{\omega CR}\right]^2\right\}} = \sqrt{\{9 + (3\tan\theta)^2\}}$$

Putting $CR = 1$ and $\tan \theta = \tan 45° = 1$ then gives

$$\omega^2 - 3\omega - 1 = 0$$

for which the valid root is

$$\omega = 3\cdot3 \text{ rad/s}$$

and

$$|A| = \sqrt{(9+9)} = 3\sqrt{2} = 4\cdot24$$

Comment

The values in this illustration, while chosen to yield a simple solution, demonstrate clearly the significance of phase-change in the amplifier itself. In a well designed phase-shift oscillator of this type, $\arg A$ would be very small, and for $\arg A = 0$, the expressions reduce to $A = 3$ and $\omega = 1/CR$. As a $10:1$ range in capacitance is obtainable with a normal variable capacitor, this form of oscillator has the advantage of frequency calibration in decades.

While the focal point of this illustration is the condition of oscillation, it is important to realise that the general theory of feedback systems is based on the assumption that A and β are both complex quantities, and that the case, though common, in which β is derived from a resistive potential divider and is wholly real, is a particular one. In this illustration β is real only at the one frequency, $\omega = 1/CR$, at which it can satisfy the threshold condition for oscillation only for $\arg A = 0$.

ILLUSTRATION 6.8

An amplifier whose intrinsic input admittance may be regarded as zero has a voltage gain $A = 3\underline{/0}$ when its high potential input and output terminals are bridged by a resistor R in series with a capacitor C. Show that the input admittance when $\omega = 1/CR$ may be exactly annulled by a resistor R in parallel with a capacitor C.

Interpretation

Let $Y_f = 1/(R-j/\omega C)$ denote the bridging (or parallel feedback) admittance. Then the effective input admittance is shown in Illustration 1.9 to be

$$Y_{in} = Y_f(1-A)$$

Substituting for Y_f and A,

$$Y_{in} = \frac{-2j\omega C}{1+j\omega CR} = -\frac{2\omega^2 C^2 R}{1+(\omega CR)^2} - j\frac{2\omega C}{1+(\omega CR)^2}$$

When $\omega = 1/CR$ this reduces to

$$Y_{in} = -\frac{1}{R} - j\frac{1}{R}$$

But at $\omega = 1/CR$, a resistor R in parallel with a capacitor C has an admittance

$$Y = \frac{1}{R} + j\omega C = \frac{1}{R} + j\frac{1}{R}$$

and thus

$$Y + Y_{in} = 0$$

Comment

The amplifier acts as a *negative impedance converter*, transforming the positive elements in the feedback path into negative ones, $G_{in} = -1/R$ and $C_{in} = -(j/R)/j\omega = -1/R\omega$, at $\omega = 1/CR$. The action of an oscillator like that in Illustration 6.7 is thus based equivalently on annulment of the negative input admittance of the amplifier by the positive admittance of the paralleled resistor and capacitor.

ILLUSTRATION 6.9

A unilateral amplifier has essentially infinite input impedance and an output resistance r_0. In response to an input voltage V_1 it delivers a short-circuit output current $Y_{21}V_1$. An oscillator is formed from this amplifier by connecting across its output terminals a parallel-tuned circuit comprising a capacitor C and an inductor L having series resistance R, which is coupled by mutual inductance M to an inductor L_0 having series resistance R_0 connected across the input terminals. Derive the threshold conditions for oscillation and the frequency of oscillation.

Discuss briefly the validity of the theory, indicating modifying factors as appropriate, when the amplifier is

(a) a valve in the common-cathode circuit;
(b) a field-effect transistor in the common-source circuit;
(c) a junction transistor in the common-emitter circuit.

Interpretation

Figure 6.8 is an equivalent circuit representing the amplifier under open-loop conditions. V_1 is an assumed excitation at the input terminals 1, 2 and V_2 is the resultant open-circuit voltage across the coil L_0 which is coupled to the tuned circuit loading the output terminals 3, 4 of the amplifier. The voltage-controlled current source $Y_{21}V_1$ is oriented arbitrarily for Y_{21} reckoned positive by the standard two-port polarity conventions, while the coils L and L_0 are oriented as indicated by the dots so that terminals 1, 3' and 2, 4' have like polarities.

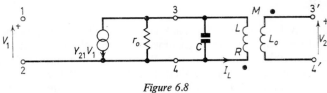

Figure 6.8

As L_0 is open-circuited, the current I_L in L is unaffected by L_0 and is simply

$$I_L = \frac{Y_{21}V_1Y_L}{\Sigma Y}$$

where $\Sigma Y = Y_L + Y_c + 1/r_0$. Then,

$$V_2 = j\omega MI_L = \frac{j\omega MY_{21}V_1}{1 + (1/r_0Y_L) + (Y_c/Y_L)}$$

V_1 is just self-sustaining when the loop is closed by linking terminals 1, 3' and 2, 4' if $V_2 = V_1$. Then,

$$1 + \frac{1}{r_0Y_L} + \frac{Y_c}{Y_L} = j\omega MY_{21}$$

or

$$1 + (R + j\omega L)/r_0 + j\omega C(R + j\omega L) = j\omega MY_{21}$$

whence, from the imaginary terms,

$$M = \frac{L}{Y_{21}r_0} + \frac{CR}{Y_{21}}$$

and from the real terms,

$$\omega^2 = \frac{1}{LC}\left[1 + \frac{R}{r_0}\right]$$

The theory is reasonably valid both for a valve and a field-effect transistor in the stated configurations, as both have substantially infinite input resistances under linear conditions. But behaviour may be seriously modified by the effects of inter-electrode capacitances. Of these, either the anode-grid or gate-drain capacitance is especially important, for not only does it pollute the desired feedback, but also provides a reverse feedback path which may invalidate the simplifying assumption of zero input admittance. (See Illustrations 1.9 and 1.12). The theory is not valid in this simple form for a junction transistor, for the base-emitter junction exhibits a low resistance (1000–1500 Ω), and the intrinsic reverse feedback cannot be ignored.

Comment

The analysis of a tuned feedback circuit may be simple, as shown, provided the amplifier can be assumed to be purely unilateral and to have a negligible input admittance. While the direct approach illustrated does not involve the explicit determination of the gain A and the feedback factor β, the reader may show that the gain from terminals 1, 2 to 3, 4 is

$$A = \frac{-Y_{21}}{Y_c + Y_L + 1/r_0}$$

and that the feedback factor β, from terminals 3, 4 to 3′, 4′, is

$\beta = -j\omega M Y_L$ (the mutual inductance is polarity-reversing)

In this particular case of an oscillator, the polarities and connections are preset for a regenerative condition. The threshold of oscillation is, however, $A\beta = +1$, which gives the same results as before. While this criterion is opposite to that derived from the standard form $A_f = A/(1 + A\beta)$, based on the polarity conventions for arbitrary series-paralleled networks shown in Figure 6.1(a), and elsewhere in a matrix context, it would obviously be pedantic to recast Figure 6.8. The polarities and connections stipulated are, in fact, consistent with the two-port chain-matrix approach in Illustration 6.10.

It is interesting to note that when $R = 0$, $Y_L = j\omega L$ and β has the wholly real value $\beta = -M/L$. The threshold conditions then reduce to

$$A = -r_0 Y_{21}, \quad M = L r_0 Y_{21}, \quad \text{and} \quad \omega^2 = 1/LC$$

and the frequency of oscillation depends solely on the tuned circuit.

ILLUSTRATION 6.10

Establish criteria for oscillation in a feedback system in terms of the chain or transfer matrices for the amplifier and feedback network when the amplifier input impedance is (a), infinite; and (b), finite. Confirm in this alternative way the threshold conditions for Illustration 6.9.

Interpretation

In the basic arrangement of Figure 6.9, $[A]_\alpha$ and $[A]_\beta$ denote the chain (transfer, $ABCD$ or a-type matrices) for the amplifier and feedback network respectively. The output current is shown as $-I_2$, reversed relative to the usual 2 port direction, in order to align it with I_1 (the chain matrix elements are, in fact, independent of the polarity convention used; see Illustration 1.21).

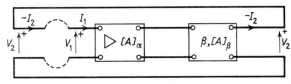

Figure 6.9

The open-loop matrix for Figure 6.9 is

$$[A]_{\alpha\beta} = [A]_\alpha \cdot [A]_\beta = \begin{bmatrix} A_\alpha A_\beta + B_\alpha C_\beta & A_\alpha B_\beta + B_\alpha D_\beta \\ C_\alpha A_\beta + D_\alpha C_\beta & C_\alpha B_\beta + D_\alpha D_\beta \end{bmatrix}$$

$$= \begin{bmatrix} A_{\alpha\beta} & B_{\alpha\beta} \\ C_{\alpha\beta} & D_{\alpha\beta} \end{bmatrix} \tag{1}$$

and the open-loop matrix equation is therefore

$$\begin{bmatrix} A_{\alpha\beta} & B_{\alpha\beta} \\ C_{\alpha\beta} & D_{\alpha\beta} \end{bmatrix} \cdot \begin{bmatrix} V_2 \\ -I_2 \end{bmatrix} = \begin{bmatrix} V_1 \\ I_1 \end{bmatrix} \tag{2}$$

(a) When the amplifier input impedance is infinite, $I_1 = 0$. Therefore when the loop is closed as indicated by the dotted links, $-I_2 = I_1 = 0$ and equation (2) reduces to

$$A_{\alpha\beta} V_2 = (A_\alpha A_\beta + B_\alpha C_\beta) V_2 = V_1 \tag{3}$$

Putting $V_2 = V_1$, the threshold of oscillation becomes

$$A_{\alpha\beta} = A_\alpha A_\beta + B_\alpha C_\beta = 1 \tag{4}$$

(b) When the amplifier input impedance is finite, I_1 is finite for an input V_1. When the loop is closed, the threshold of oscillation exists when supposed inputs V_1, I_1 to the amplifier are just reproduced as V_2, $-I_2$ at the output of the feedback network. Imposing the conditions $V_2 = V_1$ and $-I_2 = I_1$ on equation (1) gives

$$\begin{bmatrix} A_{\alpha\beta}-1 & B_{\alpha\beta} \\ C_{\alpha\beta} & D_{\alpha\beta}-1 \end{bmatrix} \cdot \begin{bmatrix} V_1 \\ I_1 \end{bmatrix} = 0 \tag{5}$$

The solutions for V_1 and I_1 are non-trivial provided the determinant of the matrix is zero, or

$$(A_{\alpha\beta}-1)(D_{\alpha\beta}-1) - B_{\alpha\beta}C_{\alpha\beta} = 0 \tag{6}$$

which is therefore the criterion for oscillation.

(c) Figure 6.8, Illustration 6.9, may be split into distinct two-port networks as shown in Figure 6.10. The amplifier is assumed to have an infinite impedance ($I_1 = 0$).

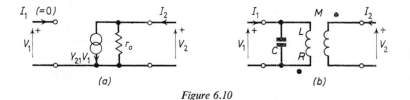

(a)   (b)

Figure 6.10

For reference, the chain-matrix equations have the general form

$$\begin{bmatrix} A & B \\ C & D \end{bmatrix} \cdot \begin{bmatrix} V_2 \\ -I_2 \end{bmatrix} = \begin{bmatrix} V_1 \\ I_1 \end{bmatrix} \quad \text{or,} \quad \begin{aligned} AV_2 - BI_2 &= V_1 \\ CV_2 - DI_2 &= I_1 \end{aligned}$$

Then, for the amplifier of Figure 6.10(a),

$$`A_\alpha = \frac{V_1}{V_2}\bigg|_{I_2=0} = -\frac{1}{Y_{21}r_0}$$

$$B_\alpha = -\frac{V_1}{-I_2}\bigg|_{V_2=0} = -\frac{1}{Y_{21}}$$

In Figure 6.10(b) let $Y_L = 1/(R+j\omega L)$, $Z_M = j\omega M$ and $Y_c = j\omega C$. Then, noting that the dotted coil ends are of like polarity,

$$V_{2(I_1=0)} = -Z_M V_1 Y_L \quad \text{and} \quad A_\beta = \frac{V_1}{V_2}\bigg|_{I_2=0} = -\frac{1}{Z_M Y_L}$$

The inductor current when $I_2 = 0$ is $I_1Y_L/(Y_c+Y_L)$ so that, noting again the dot notation,

$$V_{2(I_2 = 0)} = -Z_MY_LI_1/(Y_c+Y_L), \quad \text{and}$$

$$C_\beta = \left.\frac{I_1}{V_2}\right|_{I_2 = 0} = -\frac{Y_L+Y_c}{Z_MY_L}$$

Then, for the amplifier followed by the feedback network,

$$A_{\alpha\beta} = A_\alpha A_\beta + B_\alpha C_\beta$$

$$= \frac{1}{Z_MY_{21}}\left[1+\frac{1}{r_0Y_L}+\frac{Y_c}{Y_L}\right]$$

The closed-loop threshold condition occurs when $A_{\alpha\beta} = 1$. Then, as before,

$$1+(R+j\omega L)/r_0+j\omega C(R+j\omega L) = j\omega MY_{21}$$

Comment

There are many approaches to the threshold of oscillation in a feedback system. The chain matrix illustrated is attractive because it is systematic and leads to a general criterion, equation (6), which is readily applied to a system in which the amplifier input impedance is too low to ignore, as in the case of a junction transistor. When the input impedance is great, the criterion assumes the very simple form of equation (4) as one element ($A_{\alpha\beta}$) in the matrix product is sufficient.

ILLUSTRATION 6.11

A voltage amplifier having a substantially infinite input impedance and a low output impedance has a gain-frequency characteristic of the form

$$A(j\omega) = A(0)/(1+jK\omega)$$

where $A(0)$ is the zero frequency gain and K is a constant. Show that the application of voltage-dependent series-applied voltage feedback modifies the bandwidth, as reckoned to the half-power or 3 db frequency, by the factor $1+A(0)\beta$, where the feedback factor β may be regarded as wholly real.

Interpretation

Let $A(j\omega) = V_2(j\omega)/V_1(j\omega)$, the ratio output to input voltages. Then, for $|V_1(j\omega)|$ constant, the output power into a load is proportional to $|V_2(j\omega)|^2$, and the half-power or 3 db frequency is therefore that at which $|V_2(j\omega)|$ or $|A(j\omega)|$ falls to $1/\sqrt{2}$ of its maximum value. In this case the maximum is at $\omega = 0$, and

$$|A(j\omega)| = |A(0)/(1+jK\omega)| = A(0)/\sqrt{2}$$

when

$$|1+jK\omega| = \sqrt{2}$$

which is so when $K\omega = 1$, or $\omega = \omega_\alpha = 1/K$.

With feedback applied, the gain at any frequency ω is

$$A_f(j\omega) = \frac{Aj(\omega)}{1+A(j\omega)\beta}$$

$$= \frac{A(0)}{1+A(0)\beta+jK\omega}$$

while the zero-frequency gain is

$$A_f(0) = \frac{A(0)}{1+A(0)\beta}$$

Then,

$$\frac{A_f(j\omega)}{A_f(0)} = \frac{1+A(0)\beta}{1+A(0)\beta+jK\omega}$$

or

$$A_f(j\omega) = \frac{A_f(0)}{1+j\left[\dfrac{K\omega}{1+A(0)\beta}\right]}$$

and therefore

$$|A_f(j\omega)| = A_f(0)/\sqrt{2}$$

when

$$\frac{K\omega}{1+A(0)\beta} = \frac{\omega}{\omega_\alpha[1+A(0)\beta]} = 1$$

whence

$$\omega = \omega_{\alpha f} = \omega_\alpha[1+A(0)\beta]$$

The original 3 db frequency is thus modified by the factor $1+A(0)\beta$.

Comment

For negative feedback, as defined in general by $|1+A(j\omega)\beta(j\omega)| > 1$, and in this case by $|1+A(j\omega)\beta| > 1$, the bandwidth is increased when $1+A(0)\beta > 1$, but at the expense of gain. The maximum gain is reduced by the factor $1/[(1+A(0)\beta]$, but the output voltage could be restored to the non-feedback level without impairing the bandwidth by increasing the input voltage by the factor $1 + A(0)\beta$. The flattening of the response characteristic implied by the increased bandwidth is representative of the general linearising effects of negative feedback.

ILLUSTRATION 6.12

Figure 6.11 is a transistor amplifier incorporating a mixture of voltage-dependent and current-dependent series-applied voltage feedback, derived from the potential divider R_2, R_3 and the emitter resistor R_1. The circuit has been drawn in planar form for mesh analysis as indicated. The input and output terminals are 1, 2 and 3, 4 respectively, and the resistors have values $R_1 = 100\ \Omega$, $R_2 = 500\ \Omega$, $R_3 = 15\ \text{k}\Omega$ and $R_4 = 5\ \text{k}\Omega$. The transistor is type BC 187, and under the conditions of operation its h-parameters are $h_{ie} = 2$ kΩ, $h_{re} = 1 \cdot 4 \times 10^{-4}$, $h_{fe} = 140$ and $h_{oe} = 29$ μS. Determine the low-frequency voltage gain V_2/V_1 and the input impedance $Z_{\text{in}} = V_1/I_1$. Compare the gain with an estimate based on the formula $A_f = A/(1 + A\beta)$.

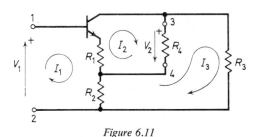

Figure 6.11

Interpretation

(a) The h-parameters must first be transformed into Z-parameters, which conform directly to mesh analysis. By reference to Illustration 1.18, $Z_{ie} = 1325$, $Z_{re} = 4 \cdot 83$, $Z_{fe} = -4 \cdot 83 \times 10^6$, $Z_{oe} = 3 \cdot 45 \times 10^4$.

The elements of the mesh-impedance matrix (for clockwise mesh currents) are

$$Z_{11} = R_1 + R_2 + Z_{ie}; \quad Z_{22} = R_1 + R_4 + Z_{oe}; \quad Z_{33} = R_2 + R_3 + R_4;$$
$$Z_{12} = -(R_1 + Z_{re}); \quad Z_{13} = Z_{31} = -R_2; \quad Z_{21} = -(R_1 + Z_{fe});$$
$$Z_{23} = Z_{32} = -R_4$$

Substituting numerical values scaled by 10^3 then gives

$$\begin{bmatrix} 1 \cdot 93 & -0 \cdot 105 & -0 \cdot 5 \\ 4830 & 39 \cdot 6 & -5 \cdot 0 \\ -0 \cdot 5 & -5 \cdot 0 & 20 \cdot 5 \end{bmatrix} \cdot \begin{bmatrix} I_1 \\ I_2 \\ I_3 \end{bmatrix} = \begin{bmatrix} 10^{-3} V_1 \\ 0 \\ 0 \end{bmatrix} \tag{1}$$

Choosing a Laplace development for $\Delta = \det [Z]$ that includes co-factors needed for finding I_2 and I_3,

$$\Delta = 1\cdot93\,\Delta_{11} - 0\cdot105\,\Delta_{12} - 0\cdot5\,\Delta_{13} = 24 \times 10^3$$

where

$$\Delta_{11} = 788, \quad \Delta_{12} = -9\cdot9 \times 10^4, \quad \Delta_{13} = -24\cdot1 \times 10^3$$

The voltage fall V_2 in the sense of I_2 is $(I_2 - I_3)R_4$. Therefore,

$$A_f = V_2/V_1 = 10^{-3}R_4(\Delta_{12} - \Delta_{13})/\Delta = -15\cdot6$$

Thus the actual sense of V_2 is the reverse of that assumed in Figure 6.11. In fact, as Δ_{12} and Δ_{13} are both negative, both I_2 and I_3 are reversed in relation to the assumed clockwise directions.

The input impedance is

$$Z_{in} = V_1/I_1 = 10^3\Delta/\Delta_{11} = 30\cdot4 \text{ k}\Omega, \text{ actual}$$

(b) By solving the equations

$$h_{ie}I_1 + h_{re}V_2 = V_1$$
$$h_{fe}I_1 + h_{oe}V_2 = I_2 = -V_2/R_4$$

(where the currents and voltages conform to the 2-port convention) the gain A of the transistor loaded with R_4 but without any feedback is found to be

$$A = V_2/V_1 = -320.$$

The feedback voltage V_{fi} across R_1 proportional to the emitter current I_e is R_1I_e. But $I_e \simeq V_2/R_4$ so that $V_{fi} \simeq (R_1/R_4)V_2$. This voltage adds in the input circuit to the voltage-dependent feedback voltage $V_{fv} = R_2/(R_2 + R_3)V_2$, so that the combined degenerative effect on gain is equivalently represented by a single negative feedback factor

$$\beta = \beta_i + \beta_v \simeq -[(R_1/R_4) + R_2/(R_2 + R_3)] \simeq -0\cdot053$$

This gives

$$A_f \simeq A/(1 + A\beta) \simeq -320/17 \simeq -17\cdot8.$$

Comment

Nodal voltage analysis is an alternative but requires 4 equations. Either way, rigorous solution is exacting because of the extremes in transistor parameter values. This is reflected in the elements of $[Z]$ in equation (1) which range from $-0\cdot105$ to 4830. Approximation has been excluded to preserve procedure, but would have been justified. Analysis with a readily approximated circuit model in place of the transistor h-matrix is yet another approach, outlined in Illustration 6.13.

The error in adapting $A_f = A/(1+A\beta)$ is only 11% in spite of the disregard of input and output loading effects (see Illustration 6.4). Precision in transistor calculations is often unjustified as the parameters available may only be nominal.

The use of both kinds of feedback in Figure 6.11 permits the output impedance Z_{out} to be controlled independently. Both forms have the same effect on gain and linearity; but voltage dependent feedback lowers Z_{out} while current dependent raises it. Z_{out} for Figure 6.11 may be calculated by finding the current traversing terminals 3, 4 in response to a voltage excitation substituted for R_4 while V_1 is replaced by the impedance (only) of any generator to be connected to terminals 1, 2.

ILLUSTRATION 6.13

Figure 6.12(a) is a circuit realisation for the h-matrix of a transistor as specified in accordance with the standard 2-port polarity convention, and Figure 6.12(b) shows an approximation to it incorporated in the feedback arrangement of Figure 6.11, Illustration 6.12. Justify the approximations and write a set of nodal-voltage equations for terminal 2 as datum. Scale the parameters in Illustration 6.12 conveniently and find V_2/V_1 and $Z_{in} = V_1/I_1$, where $V_1 = V_b$ and the voltage fall V_2 (in conformity with Figure 6.11) is $V_c - V_a$.

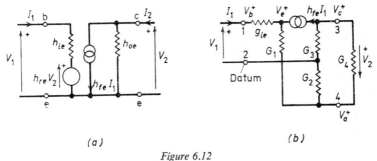

(a)   *(b)*

Figure 6.12

Interpretation

The approximations are justified because h_{oe} is small compared with the conductance of the external circuitry, and the intrinsic transistor feedback $h_{re}V_2$ is small compared with that through G_1 and G_2.

By inspection, for the assumed positive node potentials,

$$
\begin{aligned}
g_{ie}V_b - g_{ie}V_e &= I_1 \\
-g_{ie}V_b + (g_{ie}+G_1)V_e - G_1 V_a &= h_{fe}I_1 \\
-G_1 V_e + (G_1+G_2+G_4)V_a - G_4 V_c &= 0 \\
-G_4 V_a + (G_3+G_4)V_c &= -h_{fe}I_1
\end{aligned}
\tag{1}
$$

For convenience let $I_1 = 1A$. Transforming the resistances in Illustration 6.12 into conductances, scaling them (but *not* h_{fe}) by 10^4 and eliminating V_e from equations (1),

$$
\begin{aligned}
V_b - V_a &= 1{\cdot}61 \\
-10V_b + 12{\cdot}2V_a - 0{\cdot}2V_c &= -2 \\
-2V_a + 2{\cdot}67V_c &= -140
\end{aligned}
\tag{2}
$$

The scaled solutions are $V_c = -51{\cdot}1$, $V_b = 3{\cdot}38$, $V_a = 1{\cdot}77$. These give

$$
\begin{aligned}
V_2/V_1 &= (V_c - V_a)/V_b = -15{\cdot}7 \\
Z_{in_{scaled}} &= V_1/I_1 = V_b = 3{\cdot}38\ \Omega \\
Z_{in_{actual}} &= 3{\cdot}38 \times 10^4 = 33{\cdot}8\ k\Omega
\end{aligned}
$$

Comment

The gain differs negligibly from that calculated in Illustration 6.12. There is, however, an appreciable difference in Z_{in}. This is more sensitive to the omission of h_{re}; for the base-emitter resistance of the transistor, which has been taken as h_{ie} alone, is modified by the term $-h_{re}h_{fe}/(h_{oe}+G_{ce})$ where G_{ce} is the net collector-emitter load. In the actual amplifier this discrepancy is magnified by the external feedback.

Approximations to Figure 6.12(a) are widely used for the simplified analysis of transistor circuits at low frequencies for which the h-parameters are wholly real. For high frequencies the hybrid-π model (see Illustration 1.5) is the basis for approximation.

ILLUSTRATION 6.14

Figure 6.13 is the basic circuit arrangement for a class of active filter employing two feedback loops. The amplifier may be assumed to have an infinite input impedance, a negligible output impedance, and a voltage-gain A. Show that as $A \to -\infty$ the voltage transfer function becomes

$$
\frac{V_2}{V_1} = \frac{-Y_1 Y_3}{Y_3 Y_4 + Y_5(Y_1 + Y_2 + Y_3 + Y_4)}
$$

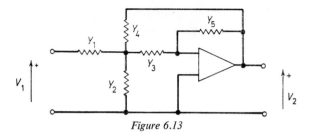

Figure 6.13

Interpretation

Let the nodes that are the junctions of Y_1, Y_2, Y_3, Y_4 and Y_3, Y_5 be respectively at algebraically positive potentials V_3 and V_4. Transforming V_1 in series with Y_1 into a current source Y_1V_1 in parallel with Y_1 and taking the common input-output line as datum,

$$(Y_1+Y_2+Y_3+Y_4)V_3 \quad -Y_3V_4-Y_4V_2 \quad = Y_1V_1 \tag{1}$$

$$-Y_3V_3+(Y_3+Y_5)V_4-Y_5V_2 \quad = 0 \tag{2}$$

$$-Y_4V_3 \quad -Y_5V_4+(Y_4+Y_5)V_2 = 0 \tag{3}$$

Substituting V_2/A for V_4 then gives

$$(Y_1+Y_2+Y_3+Y_4)V_3-(Y_4+Y_3/A)V_2 = Y_1V_1 \tag{4}$$

$$-AY_3V_3+(Y_3+Y_5-AY_5)V_2 = 0 \tag{5}$$

In the limit as $A \rightarrow -\infty$ these become

$$(Y_1+Y_2+Y_3+Y_4)V_3-Y_4V_2 = Y_1V_1 \tag{6}$$

$$V_3 = -V_2Y_5/Y_3 \tag{7}$$

from which the expression given for V_2/V_1 is easily obtained.

Comment

Special feedback formulae are useful only for relatively simple arrangements. For example, equation (7) could have been predicted by recognising that the amplifier together with Y_3 and Y_5 constitutes an operational amplifier with an established gain property (see Illustration 1.10). Direct loop current or a nodal-voltage analysis is often the best approach, and in this example the transfer function is obtained very easily by the nodal-voltage method. This method is usually preferable to loop-current analysis when the network incorporates an amplifier whose input impedance is to be assumed infinite;

for while the input terminals are readily assigned node potentials, the terminal-pair does not present a closed path for the circulation of an assumed loop current.

The potential of Figure 6.13 as a filter is exemplified by a circuit in which Y_1, Y_3 and Y_4 are the admittances of capacitors C_1, C_3 and C_4, while Y_2 and Y_5 are conductances G_2 and G_5. Then, in terms of the general frequency variable s,

$$\frac{V_2(s)}{V_1(s)} = \frac{-s^2 C_1 C_3}{s^2 + C_3 C_4 + s G_5 (C_1 + C_3 + C_4) + G_2 G_5}$$

For $s = j\omega$ this has the form

$$\frac{V_2(j\omega)}{V_1(j\omega)} = \frac{K\omega^2}{(c - a\omega)^2 + jb\omega}$$

or

$$\left| \frac{V_2(j\omega)}{V_1(j\omega)} \right|^2 = \frac{K}{\left(\dfrac{c}{\omega^2} - a \right)^2 + \left(\dfrac{b}{\omega} \right)^2}$$

which is a high-pass characteristic.

Bibliography

1. DESOER, CHARLES A. and KUH, ERNEST S. *Basic Circuit Theory*, McGraw-Hill (1969). International Student Edition available.
2. GUILLEMAN, ERNST A. *Introductory Circuit Theory*, John Wiley & Sons Inc., New York, and Chapman & Hall, London (1953).
3. LAGASSE, J. *Study of Electric Circuits*, Butterworth (1965).
4. LAGASSE, J. *Linear Circuit Theory*, Butterworth (1968).
5. ROGERS, F. E. *Topology & Matrices in the Solution of Networks*, Butterworth (1965).
6. STEWART, J. L. *Circuit Theory and Design*, John Wiley & Sons Inc., New York, and Chapman and Hall, London (1956).
7. SMITH, R. J. *Circuits, Devices and Systems*, John Wiley & Sons Inc., New York, and Chapman and Hall, London (1966).
8. CHENG, D. K. *Analysis of Linear Systems*, Addison Wesley Publishing Co. (1959).
9. PIPES, LOUIS A. *Applied Mathematics for Engineers and Physicists*, McGraw-Hill (1958).
10. SPENCE, ROBERT. *Linear Active Networks*, Wiley-Interscience (1970).
11. HAYKIN, S. S. *Junction Transistor Circuit Analysis*, Butterworth (1962).
12. HAYKIN, S. S. and BARRETT, R. *Transistor Circuits in Electronics*, 2nd ed., Butterworth (1971).
13. MILLMAN, J. and HACLIAS, C. G. *Electronic Devices and Circuits*, McGraw-Hill (1967).
14. THORNTON, R. D. *et al. Multistage Transistor Circuits*, SSEC, Vol. 5., John Wiley & Sons Inc. (1965).
15. JOHNS, P. B. and ROWBOTHAM T. R. *Communication Systems Analysis*, Butterworth (1972).

Index